EAI/Springer Innovations in Communication and Computing

Series Editor

Imrich Chlamtac, European Alliance for Innovation, Ghent, Belgium

Editor's Note

The impact of information technologies is creating a new world yet not fully understood. The extent and speed of economic, life style and social changes already perceived in everyday life is hard to estimate without understanding the technological driving forces behind it. This series presents contributed volumes featuring the latest research and development in the various information engineering technologies that play a key role in this process.

The range of topics, focusing primarily on communications and computing engineering include, but are not limited to, wireless networks; mobile communication; design and learning; gaming; interaction; e-health and pervasive healthcare; energy management; smart grids; internet of things; cognitive radio networks; computation; cloud computing; ubiquitous connectivity, and in mode general smart living, smart cities, Internet of Things and more. The series publishes a combination of expanded papers selected from hosted and sponsored European Alliance for Innovation (EAI) conferences that present cutting edge, global research as well as provide new perspectives on traditional related engineering fields. This content, complemented with open calls for contribution of book titles and individual chapters, together maintain Springer's and EAI's high standards of academic excellence. The audience for the books consists of researchers, industry professionals, advanced level students as well as practitioners in related fields of activity include information and communication specialists, security experts, economists, urban planners, doctors, and in general representatives in all those walks of life affected ad contributing to the information revolution.

Indexing: This series is indexed in Scopus, Ei Compendex, and zbMATH.

About EAI

EAI is a grassroots member organization initiated through cooperation between businesses, public, private and government organizations to address the global challenges of Europe's future competitiveness and link the European Research community with its counterparts around the globe. EAI reaches out to hundreds of thousands of individual subscribers on all continents and collaborates with an institutional member base including Fortune 500 companies, government organizations, and educational institutions, provide a free research and innovation platform.

Through its open free membership model EAI promotes a new research and innovation culture based on collaboration, connectivity and recognition of excellence by community.

More information about this series at http://www.springer.com/series/15427

S. Velliangiri • M. Gunasekaran • P. Karthikeyan
Editors

Secure Communication for 5G and IoT Networks

EAI
RESEARCH MEETS INNOVATION

Springer

Editors
S. Velliangiri
CSE B V Raju Institute of Technology
Narasapur, India

P. Karthikeyan
CSE Jain (Deemed-to-be University)
Bengaluru, India

M. Gunasekaran
Electrical Engineering Technology
University of California
Davis, CA, USA

ISSN 2522-8595 ISSN 2522-8609 (electronic)
EAI/Springer Innovations in Communication and Computing
ISBN 978-3-030-79765-2 ISBN 978-3-030-79766-9 (eBook)
https://doi.org/10.1007/978-3-030-79766-9

This Springer imprint is published by the registered company Springer Nature Switzerland AG
The registered company address is: Gewerbestrasse 11, 6330 Cham, Switzerland

Preface

This book highlights state-of-the-art research on secure communication of 5G and the Internet of Things (IoT) Networks, along with related areas to ensure secure and Internet-compatible IoT systems. It discusses 5G and IoT security and privacy challenges and energy-efficient approaches to improve the ecosystems through communication. The book addresses the secure communication and privacy of the 5G and IoT technologies while also revealing the impact of IoT technologies on several scenarios in industrial automation. Intended as a comprehensive introduction, it offers an in-depth analysis. It provides scientists, engineers and professionals with the latest techniques, frameworks and strategies used in 5G and IoT technologies. The book includes intelligent solutions by utilizing artificial intelligence and machine learning to improve the performance of the 5G and IoT security protocols and models. Additionally, this book discusses various applications that use 5G and IoT in multimedia data communications, healthcare and energy.

Narasapur, India	S. Velliangiri
Davis, CA, USA	M. Gunasekaran
Bengaluru, India	P. Karthikeyan

Preface

Contents

Chapter 1
Security Challenges in 5G and IoT Networks: A Review

G. Edwin Prem Kumar, M. Lydia, and Yoash Levron

1.1 Introduction

The fifth-generation (5G) networks are destined to emerge as one of the chief drivers of applications and devices involving the Internet of Things (IoT). Wireless technologies have steadily grown from 2G networks supporting only voice, 3G networks for both voice and data, and 4G for broadband connectivity. 4G networks powered by Long Term Evolution (LTE), outperformed all other rival technologies like Bluetooth, Zigbee, Long Range Radio (Lo-Ra) technologies. The challenges faced in 4G in terms of speed, disruptions, and reliability led to the evolution of 5G networks. IoT applications that had already started mushrooming with 3G and 4G networks have set to become the greatest beneficiary of 5G networks. The 5G-based IoT will be able to connect billions of IoT devices and facilitate the implementation of advanced IoT applications like Smart cities, Smart homes, Smart grids, and so on, with this very high-speed connectivity and reliability. The 5G-IoT architecture is expected to be well-coordinated, agile, intelligent, and fully automatic [1]. It would provide enormous number of connections of varying standards and would also implement network functions on-demand, using cloud-based Radio Access Network (RAN). The requirements of 5G-IoT include high data rate, scalability,

G. E. P. Kumar (✉)
Department of Information Technology, Sri Krishna College of Engineering and Technology, Coimbatore, Tamil Nadu, India

M. Lydia
Department of Mechatronics Engineering, Sri Krishna College of Engineering and Technology, Coimbatore, Tamil Nadu, India

Y. Levron
The Viterbi Faculty of Electrical Engineering, Technion-Israel Institute of Technology, Haifa, Israel
e-mail: yoashl@ee.technion.ac.il

S. Velliangiri et al. (eds.), *Secure Communication for 5G and IoT Networks*, EAI/Springer Innovations in Communication and Computing,
https://doi.org/10.1007/978-3-030-79766-9_1

low latency, mobility, battery life, and security. Network function virtualization (NFV), software-defined wireless sensor network (SD-WSN), cognitive radio (CR), and device-to-device (D2D) communications are some of the key enablers in 5G-IoT architecture [2].

The IoT platform comprises of intelligent, heterogeneous devices connected through the internet. The layered architecture of IoT comprises of perception layer, network layer, and application layer. The most common security attacks in the perception layer are interference from radio frequency (RF), jamming of nodes, false node injection, sleep deprivation, and tampering of hardware. Man-in-the-middle attacks, spoofing, sinkhole, and Sybil attacks are common in the network layer and Denial-of-service (DoS), malicious script and phishing attacks take place in the application layer [3]. Insufficient authorization procedures, insecure network services and interfaces, inadequate security configuration, and unreliable communication platforms are some of the attack vectors in IoT systems [4]. Khattak et al. critically reviewed the attacks in the perception layer, which includes attacks in radio frequency identification (RFID), WSN, and RFID sensor network (RSN) [5]. Securing the physical objects in IoT networks has also become an interesting area of research. The physical objects in IoT networks are unpredictable in the spatial and temporal realm, have multiple identities, are resource-constrained, dynamic, highly heterogeneous, and socially aware [6]. These features make securing the IoT product life cycle a big challenging task. An overview of security mechanisms spanning the entire life cycle of IoT products is presented in [7].

As the quantum of connectivity, devices, and coverage is set to reach massive high, serious concerns regarding the security and privacy of the users, devices, and data involved have arisen. The security concerns in 5G-IoT networks must vigilantly consider sealing the cracks and securing weak links in the authentication mechanism, assurance, storage, mobility, and compatibility [1]. Security needs to be ensured both during design and operations, by means of security mechanism, suitable network architectures, and well-tested protocols [8]. Higgins points out that with the integration of 5G and IoT, the risk associated with the innovation is inevitable [9]. Trust levels need to be identified as an important entity in securing the privacy of 5G and IoT networks. Trust models tailor-made for 5G networks, incorporating network slice trust degree concept and facilitating ubiquitous computing is the need of the hour. In a 5G-IoT scenario, challenges like Quality-of-Service, mobility, and slicing stand to gain greater advantage along with fog computing. Sicari et al. concluded that security and privacy constraints in the 5G-IoT scenario need to address issues pertaining to integrity, confidentiality and non-repudiation, intrusion detection, authentication, and access control, key management, trust, and enforcement of policies [10].

With sufficient light thrown on the need for securing the 5G-IoT networks, we continue to discover the technologies for securing the 5G-IoT scenario, the research work carried out on securing the 5G-IoT applications, and the research challenges and avenues to be explored in the future.

1.2 Technologies for Securing 5G-Based IoT Networks

Detection of points, contextual and collective anomalies, network intrusion, malware, ransomware, and intruders are the identified application areas of IoT security. The technologies explored for providing security in the 5G-IoT scenario are presented here (Fig. 1.1).

1.2.1 Authentication and Access Control

Authentication is rendered to be the prime security mechanism in IoT applications. However, these mechanisms have been challenged by attacks like node impersonation, node bypassing, and node capture attacks. Access control involves identification, authentication, authorization, and accountability [5]. Behrad et al. proposed a slice-specific authentication and access control (SSAAC) protocol for a 5G-IoT scenario using three network functions, namely Gateway virtual function provided by a third party, Gateway function repository, and radio resource control connection endpoint [11]. The security aspect of the newly designed RAN was compared with the conventional RAN. Pesic et al. developed the CAAVI-RICS model for securing edge computing platforms [12]. They claimed that the proposed methodology could be used to model most of the attacks in the cyber-physical system. The Credibility,

Fig. 1.1 Technologies for securing the 5G-IoT framework

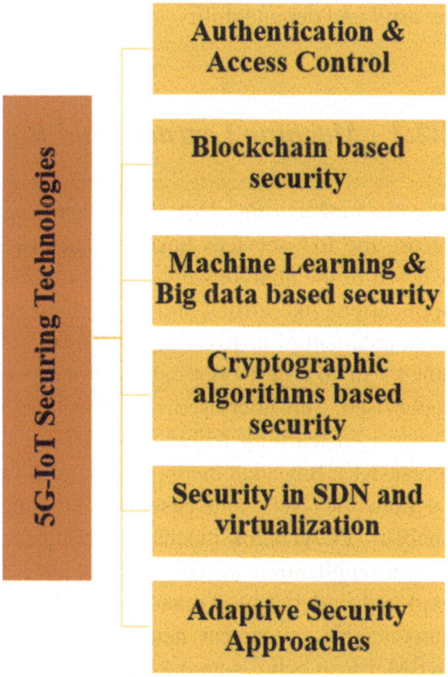

Authentication, Authorization, Verification, and Integrity (CAAVI) model security mechanisms in edge computing platforms would be robust, responsive, dynamic, effective, and resource preserving.

Wazid et al. proposed a lightweight authentication mechanism for a cloud-based IoT environment [13]. The protocol was simulated in NS2 and its effectiveness over other prevailing schemes was analyzed. Patel and Doshi presented a message queueing telemetry transport (MQTT)-based framework for providing validation and access control [14].

1.2.2 Blockchain-Based Security

Adoption of blockchain technology in this 5G-IoT scenario, which usually includes a trusted intermediary, clearly enhances the robustness and trustworthiness of the data involved. A block basically contains the data on transactions, timestamp, cryptographic hash function, reference to the previous block, and smart contracts if required [10]. Blockchain technology provides security based on the transparency of data and auditability, information distribution, robustness, and decentralization. Haris and Al-Madeed presented an exhaustive review on the integration of blockchain technology in a 5G-IoT environment [15]. Qian et al. proposed a novel sophisticated security management algorithm based on blockchain for different layers of IoT architecture [16]. A device identification-based key algorithm was developed based on the interaction between the IoT devices and blockchain database to ensure security and dependability.

1.2.3 Machine Learning and Big Data–Based Security

The machine learning (ML), deep learning (DL), and big data technologies used to secure the IoT services have also been reviewed here. Edge computing platform ushered in the analytics revolution and also enhances the performance of complex applications like virtual and augmented reality, smart city applications, and so on. Establishment of a strong defensive framework and detection of active and passive attacks in IoT networks can be done based on ML techniques, which comprise supervised, unsupervised, and reinforcement learning methodologies [17]. Supervised learning techniques include support vector machine (SVM), random forest (RF), Bayesian theorem, k-nearest neighbor (k-NN), decision tree (DT), neural network (NN), and ensemble learning. Techniques like principal component analysis (PCA) and k-means clustering comprise the unsupervised techniques.

The application of DL and big data technologies for IoT security has been explored in detail by Amanullah et al. [18]. Deep learning architectures including autoencoders, recurrent neural networks (RNN), restricted Boltzmann machine (RBM), deep belief network (DBN), long short-term memory (LSTM), convolution neural networks (CNN), and so on, can be used for IoT security. Big data technologies like Apache Spark, Apache Hadoop, and Apache Storm along with DL techniques have been used in securing the 5G-IoT net-works especially in the detection of anomalies, node abnormal behavior, and real-time intrusion. He et al. developed an extremely dependable, deep CNN-based multiuser, multi-input multi-output detector for 5G and beyond 5G enabled IoT [19]. The detection performance was increased by incorporating user selection and scheduling criteria. Thantharate et al. proposed a network slicing function based on DL for securing the 5G and beyond 5G networks [20]. The offering of network slicing-as-a-service performed competently with advanced security and dependability, thus resulting in a 'Secure5G' network. Al-Turjman explored the possibility of using big data-based analytics in securing the Internet of Nano-things (IoNT) [21].

1.2.4 *Cryptographic Algorithms-Based Security*

Traditional cryptographic-based security is highly complex and resource consuming. Hence Kumar et al. developed a lightweight signcryption technique for securing the perception layer in IoT [22]. The proposed scheme used a lightweight hash technique and was proved to be robust and resilient against replay attack, man-in-the-middle attack, impersonation, and device attacks. Dey et al. proposed a methodology based on physical layer key generation and encryption as safety measures [23]. They also explored the efficacy of advanced encryption standard (AES) and compared it with other methods and concluded that the AES algorithm outperformed the other algorithms. El-Latif et al. proposed two effective quantum-based hash functions and security protocols for protecting data in 5G networks [24]. The proposed quantum-based security algorithm outperformed the existing methods and proved to be robust against a man-in-the-middle attack, measurement attack, and message attack.

1.2.5 *Security in Software-Defined Framework and Virtualization*

5G-IoT scenario is largely based on software-defined network (SDN) and network function virtualization (NFV). Sethi et al. proposed a secure self-optimized SDN framework for narrow band-IoT in 5G networks [25]. An exhaustive review on the security framework in virtualized 5G networks was carried out by Suraci et al. [26]. The security risks and threats underlying the 5G landscape were analyzed from the stakeholders' point of view. Shafi et al. analyzed the security threats in novel technologies incorporated in 5G-IoT scenarios like device-to-device communication, spectrum sharing, ultra-dense networks, and so on [27]. They also analyzed the 5G–New Radio security issues from the aspect of artificial rain and dust.

1.2.6 Adaptive Security Approaches

In resource-constrained networks, adaptive security approaches have proven to be successful to offer energy-efficient security. These approaches are generally classified into data-centric approaches and threat-centric approaches. Data-centric approaches examine the sensitivity of the data involved and adapt the security methodology whereas the threat-centric approach examines the threats involves and adapts accordingly. Hellaoui et al. proposed an end-to-end adaptive approach for an energy-efficient solution in securing the 5G-based IoT [28].

1.3 Securing the 5G-Based IoT Applications

This section presents a brief overview of the various 5G-based IoT applications and the security mechanisms incorporated in these (Fig. 1.2).

1.3.1 Securing the Smart Grid

Smart Grid (SG) is chiefly an automated power delivery network, with improved protection, reliability, and efficient monitoring. Hui et al. conducted an exhaustive survey on the 5G-IoT scenario for demand response in SG networks. Application of

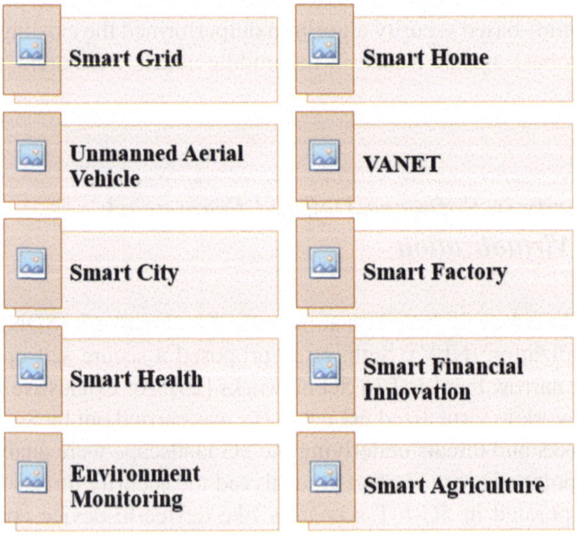

Fig. 1.2 Securing the 5G-IoT applications

5G-IoT aids in robust security, high-speed transfer of data, massive increase in connection of flexible loads, and enhanced reliability in SG networks [29]. Borgaonkar and Jaatun analyzed the implementation of secure IoT in SG networks [30]. They presented the benefits of 5G in two types of IoT devices, namely regular IoT devices like drones, connected cars, and so on, and small resource inhibited remotely placed sensors and actuators. They concluded by formulating technical principles to be adhered to for ensuring layer-based security in SG networks.

1.3.2 Securing the Smart Home

Smart homes promise increased comfort and a sophisticated lifestyle. However, the issue of security and problems of cybersecurity remains a serious undeniable threat. Wasicek has pointed out that micro-segmentation in 5G smart homes is a promising area to be explored to ensure security [31]. Jha et al. proposed layer-wise security for smart home applications [32]. They mathematically modeled the security of a smart home system, assessed the secrecy capacity and outage probability to tackle layer-wise security attacks.

1.3.3 Securing the Intelligent UAV

Unmanned aerial vehicles (UAV) have found application in several areas. Nomikos et al. proposed a mobile 5G RAN based on UAV to enable immense connectivity between IoT devices and mobile users. They analyzed the security threats in flying ad-hoc networks (FANETs) and concluded that independent behavior, self-organization, anonymity, and lack of fixed infrastructure are among the chief causes that affect its security [33]. Wang et al. proposed a novel methodology to guarantee secure and effective data aggregation in UAVs [34]. The proposed secure methodology known as the Intelligent UAV-based Data Aggregation algorithm (IDAA) outperformed the conventional techniques.

1.3.4 Securing the VANET

Intelligent transportation system has been greatly benefitted by the 5G-IoT scenario. However, the security challenges faced by Vehicular Ad-hoc Networks (VANETs) have also multiplied. Xie et al. proposed a trustworthy solution for 5G-VANETs based on blockchain technology [35]. The efficacy of the proposed methodology was validated with regard to user security, malicious vehicle analysis, and compromised roadside units. Al-Turjman and Lemayian presented a comprehensive overview on securing the VANETs [36]. They categorized the attacks on vehicle sensor networks as physical attacks, side-channel attacks, environmental attacks, cryptoanalysis attacks, software, and network attacks. Hussain et al. presented a review on the amalgamation VANETs and 5G security [37]. According to them, the 5G threat landscape included 5G core network protection, 5G cloud RAN protection, threats in the device or the end-user, and threats in business application.

1.3.5 Securing the Smart City

5G and IoT infrastructures are definitely a great asset for the reliable functioning of smart cities. Securing the smart city application is also a challenging task. Serrano proposed a combination of blockchain techniques with random neural networks for ensuring cybersecurity in the 5G-IoT networks [38]. The developed model was proved to provide security to all the layers of the 5G networks and node authentication in smart buildings. Godquin et al. developed a methodology to optimally position security solutions using graph theory [39]. The proposed technique was validated for smart city infrastructure.

1.3.6 Securing the Smart Factory

The penetration of 5G-IoT networks has revolutionized industries and the manufacturing process. It has resulted in smart interconnection of shop floors, application of big data analytics, cloud computing, virtual and augmented reality, and smart manufacturing. Cheng et al. analyzed the security requirements in 5G integrated smart manufacturing [40]. The application of blockchain technique in securing the 5G-enabled IoT in industries was put forth by Mistry et al. [41]. They have presented a comprehensive review and analyzed the problems and solutions in smart factories.

1.3.7 Securing the Smart Health

The healthcare industry has witnessed a tremendous change with the advent of 5G wireless networks and IoT. There has been an evolution in the development of scalable, dependable, and secure protocols and architectures for data acquisition, transmission, and storage from body sensors [42]. 5G-IoT networks have been a great boon to Sports Health monitoring as well [43]. Minahil et al. proposed an elliptic curve cryptography-based authentication protocol for e-health clouds [44]. The methodology was lightweight and was proved to outperform conventional techniques in terms of computational complexity and robustness.

1.3.8 Security in Other Applications

Financial Innovation

5G-IoT networks are poised to impact the financial sector as well. Cheng has pointed out that secure protocols can pave the way to noteworthy trade cycles and reduce transaction delays [45]. It can lay the platform for innovation in smart rural finance.

Environment

5G-enabled IoT networks play an important role in environmental aspects too. Han et al. proposed a blockchain-based secure architecture for monitoring and measurement of air pollution [46].

Smart Agriculture

5G-based IoT networks play a critical role in smart and precision farming too. Tang et al. presented an exhaustive survey on the role of 5G networks in agriculture and the need for a secure communication channel, connectivity, and storage [47].

1.4 Research Challenges in Securing 5G-IoT Networks

The research works pertaining to technologies involved in securing 5G-IoT networks have been summarized in Table 1.1. The various issues that continue to challenge researchers in the area of 5G-IoT security have been presented in this section.

1.4.1 Challenges Related to 5G-IoT Architecture

Scalability, heterogeneity, and interoperability continue to remain a challenge in 5G-IoT architecture. Standardization has gained much significance as 5G-IoT has evolved as a very complex ecosystem. The need to conform to both technology and regulatory standards is a challenging task yet to be addressed. There is an increasing need for the development of novel transmission techniques based on the distance

Table 1.1 Securing technologies applicable to 5G-IoT—a summary

Securing technology	Proposed mechanism
Authentication and Access Control (AAC)	Slice Specific AAC [11]
	CAAVI-RICS [12]
	Lightweight Authentication Mechanism [13]
	Novel MQTT security framework [14]
Blockchain-based security	Decentralized IoT security [16]
Machine Learning and Big data-based security	DL-based ultrareliable multiuser, multi-input, multi-output detector [19] DL-based model for robust network slicing framework [20]
Cryptographic algorithms based security	Lightweight signcryption method [22] Physical layer key generation and encryption [23] Quantum-based security protocol [24]
Security in SDN and virtualization	5G SoftAir architecture
Adaptive security	Coalition game algorithm with trust evaluation [28]

and bandwidth of transmission. The impact of path loss and noise needs to be analyzed meticulously.

1.4.2 Challenges Related to Blockchain Technologies

Block mining in Blockchain technology consumes a lot of energy and is computationally complex. This technology is also memory intensive and power consuming.

Maintaining security and privacy in every smart application is a major open research challenge. Preserving confidentiality, integrity, and authenticity of data and users need to be ensured by the incorporation of suitable cryptographic algorithms or trust levels. Development of compatible protocols for securing the IoNT is also an area to be explored. Security in ubiquitous data aggregation and visualization, state estimation, and distribution networks in SG networks is a required area of research. In VANETS, secure positioning, secure data acquisition, and verification, secure routing and delay have emerged as potential areas of research.

1.4.3 Challenges Related to Energy

Deficiency in standardization, increase in energy consumption and lower storage capacity and computational power are some of the challenges that need to be explored in 5G-enabled IoT applications. Energy-efficient solutions and viable NFV are the need of the hour. Research in energy-efficient protocols and spectrum harvesting techniques is a key trend yet to be explored in a big way. Solutions that are context-aware and convergent are needed.

1.4.4 Challenges Related to Artificial Intelligence

Artificial Intelligence (AI) based solutions for abnormal traffic monitoring is an emerging research area in 5G-IoT networks. Designing an effective traffic detection solution in IoT is a real challenge. Diverse devices in IoT have made identification a major issue. AI- and ML-based solutions of identity verification are a major challenge in 5G-IoT networks. Integration of AI with blockchain technology is also an open area of research. Authentication of training datasets is a significant challenge in ML-based security solutions.

1.4.5 Challenges Related to Software Defined Networks

The efficacy of software-defined framework in the 5G-IoT scenario needs to be ascertained. Securing SDN is also a research challenge. The trade-off between the performance of the SDN and its security poses a significant test. Evolving hybrid SDN control architecture for 5G-IoT networks is also an emerging research area.

1.5 Conclusion

Accurate detection of security threats, optimum computational and time complexity, reduced energy constraint, and security at edge computing platforms are the chief requirements of any security algorithm. This chapter presents an exhaustive review on authentication and access control protocols, key management, trust management, and intrusion detection systems for 5G and IoT networks. The role of edge computing and blockchain technology in incorporating security to 5G networks has also been explored. Since security and privacy methodologies are generally complex and resource-intensive, energy-efficient, effective and lightweight compatible solutions are the need of the hour. 5G will definitely be a great enabler in making the IoT smarter and ubiquitous. The research challenges presented give a better and in-depth insight into the need for securing 5G-IoT networks.

References

1. Li S, Xu LD, Zhao S (2018) 5G Internet of Things: A Survey. J. of Indus. Inform. Integ. 10: 1–9.
2. Shafique K, Khawaja BA, Sabir F, Qazi S, Mustaqim M (2020) Internet of Things (IoT) for Next-Generation Smart Systems: A Review of Current Challenges, Future Trends and Prospects for Emerging 5G-IoT Scenarios. IEEE Access 8: 23022–23040.
3. Yugha R, Chithra S (2020) A survey on technologies and security protocols: Reference for future generation IoT. J. of Network and Computer Applications 169: 102763.

4. Noor MBM, Hassan WH (2019) Current research on Internet of Things (IoT) security: A survey. Computer Networks 148: 283–294.
5. Khattak HA, Shah MA, Khan S, Ali I, Imran M (2019) Perception layer security in Internet of Things. Future Generation Computer Systems 100: 144–164.
6. Yao X, Farha F, Li R, Psychoula I, Chen L, Ning H (2020) Security and privacy issues of physical objects in the IoT: Challenges and opportunities. Digital Communications and Networks. In Press (Available Online)
7. Yousefnezhad N, Malhi A, Framling K (2020) Security in product lifecycle of IoT devices: A survey. J. of Network and Computer Applications 171: 102779.
8. Mazurczyk W, Bisson P, Jover RP, Nakao K, Cabaj K (2020) Special issue on Advancements in 5G Networks Security, Future Generation Comp. Sys. 110: 314–316.
9. Higgins D (2020) Innovation and risk walk hand-in-hand with 5G and IoT. Network Security, 16–18.
10. Sicari S, Rizzardi A, Coen-Porisini A (2020) 5G in the internet of things era: An overview on security and privacy challenges. Computer Networks 179: 107345.
11. Behrad S, Bertin E, Tuffin S, Crespi N (2020) A new scalable authentication and access control mechanism for 5G-based IoT. Future Generation Comp. Sys. 108: 46–61.
12. Pesic S, Ivanovic M, Radovanovic M, Badica C (2020) CAAVI-RICS model for observing the security of distributed IoT and edge computing systems. Simulation Modelling Practice and Theory 105: 102125.
13. Wazid M, Das AK, Bhat V, Vasilakos AV (2020) LAM-CIoT: Lightweight authentication mechanism in cloud-based IoT environment. J. Network and Computer Applications 150: 102496.
14. Patel C, Doshi N (2020) A novel MQTT security framework in generic IoT model. Procedia Computer Science 171: 1399–1408.
15. Haris RM, Al-Maadeed S (2020) Integrating Blockchain Technology in 5G enabled IoT: A Review. IEEE Intl. Conf. on Informatics, IoT and Enabling Technologies, 367–371.
16. Qian Y, Jiang Y, Chen J, Zhang Y, Song J, Zhou M, Pustisek M (2018) Towards decentralized IoT security enhancement: A block-chain approach. Computers and Electrical Engg. 72: 266–273
17. Tahsien SM, Karimipour H, Spachos P (2020) Machine learning based solutions for security of Internet of Things (IoT): A survey. J. of Network and Computer Applications 161: 102630.
18. Amanullah MA, Habeeb RAA, Nasaruddin FH, Gani A, Ah-med E, Nainar ASM, Akim NM, Imran M (2020) Deep learning and big data technologies for IoT security. Computer Communications 151: 495–517.
19. He K, Wang Z, Li D, Zhu F, Fan L (2020) Ultra-reliable MU-MIMO detector based on deep learning for 5G/B5G-enabled IoT. Physical Communication 43: 101181.
20. Thantharate A, Paropkari R, Walunj V, Beard C, Kankariya P (2020) Secure5G: A Deep Learning Framework Towards a Secure Network Slicing in 5G and Beyond. 10th Annual Computing and Communication Workshop and Conference, 0852-0857
21. Al-Turjman F (2020) Intelligence and security in big 5G-oriented IoNT: An overview. Future Generation Computer Systems 102: 357–368.
22. Kumar A, Saha R, Alazab M, Kumar G (2020) A Lightweight Signcryption Method for Perception Layer in Internet-of-Things. J. of Information Security and Applications 55: 102662.
23. Dey A, Nandi S, Sarkar M (2018) Security measures in IoT based 5G networks. 3rd International Conference on Inventive Computation Technologies, pp. 561–566.
24. El-Latif AAA, Abd-El-Atty B, Andraca SEV, Mazurczyk W (2019) Efficient quantum-based security protocols for information sharing and data protection in 5G networks. Future Generation Computer Systems 100: 893–906.
25. Sethi A, Jain SK, Vijay S (2020) Secure self-optimizing soft-ware defined framework for NB-IoT toward 5G. Procedia Computer Science 171: 2740–2749
26. Suraci C, Araniti G, Abrardo A, Bianchi G, Iera A (2021) A stakeholder-oriented security analysis in virtualized 5G cellular networks. Computer Networks 184: 107604.

27. Shafi M, Jha RK, Sabraj M (2020) A survey on security issues of 5G NR: Perspective of artificial dust and artificial rain. J. of Network and Computer Applications 160: 102597.
28. Hellaoui H, Koudil M, Bouabdallah A (2020) Energy Efficiency in Security of 5G-Based IoT: An End-to-End Adaptive Approach. IEEE Internet of Things Journal 7 (7): 6589–6602.
29. Hui H, Ding Y, Shi Q, Li F, Song Y, Yan J (2020) 5G network-based Internet of Things for demand response in smart grid: A survey on application potential. Applied Energy 257: 113972.
30. Borgaonkar R, Jaatun MG (2019) 5G as an enabler for secure IoT in the Smart Grid. First Intl. Conf. on Societal Automation, 1–7.
31. Wasicek A (2020) The future of 5G smart home network security is micro-segmentation. Network Security 11: 11–13
32. Jha RK, Puja, Kour H, Kumar M, Jain S (2021) Layer based security in Narrow Band In-ternet of Things (NB-IoT). Computer Networks 185: 107592.
33. Nomikos N, Michailidis ET, Trakadas P, Vouyioukas D, Karl H, Martrat J, Zahariadis T, Papadopoulos K, Voliotis S (2020) A UAV-based moving 5G RAN for massive connectivity of mobile users and IoT devices. Vehicular Communications 25: 100250.
34. Wang X, Garg S, Lin H, Kaddoum G, Hu J, Alhamid MF (2021) An Intelligent UAV based Data Aggregation Algorithm for 5G-enabled Internet of Things. Computer Networks 185: 107628.
35. Xie L, Ding Y, Yang H, Wang X (2019) Blockchain-Based Secure and Trustworthy Internet of Things in SDN-Enabled 5G-VANETs. IEEE Access 7: 56656–56666
36. Al-Turjman F, Lemayian JP (2020) Intelligence, security, and vehicular sensor networks in internet of things (IoT)-enabled smart-cities: An overview. Computers and Electrical Engineering 87: 106776.
37. Hussain R, Hussain F, Zeadally S (2019) Integration of VANET and 5G Security: A review of design and implementation issues. Future Generation Computer Systems 101: 843–864.
38. Serrano W (2021) The Blockchain Random Neural Network for cybersecure IoT and 5G infrastructure in Smart Cities. J. of Network and Computer Applications 175: 102909.
39. Godquin T, Barbier M, Gaber C, Grimault JL, Bars JML (2020) Applied graph theory to security: A qualitative placement of security solutions within IoT networks. J. Information Security and Ap-plications 55: 102640
40. Cheng J, Chen W, Tao F, Lin CL (2018) Industrial IoT in 5G environment towards smart manufacturing. J. of Indus. Information Integ. 10: 10–19.
41. Mistry I, Tanwar S, Tyagi S, Kumar N (2020) Blockchain for 5G-enabled IoT for industrial automation: A systematic review, solutions, and challenges. Mechanical Systems and Signal Processing 135: 106382.
42. Rodrigues JJPC, Zeadally S, Kumar N (2017) Special Issue on 5G Wireless Networks for IoT and Body Sensors. Computer Networks 129: 335–339.
43. Zhan K (2021) Sports and Health Big Data System Based on 5G Network and Internet of Things System. Microprocessors and Microsystems 80: 103363.
44. Minahil, Ayub MF, Mahmood K, Kumari S, Sangaiah AK (2021) Lightweight authentication protocol for e-health clouds in IoT based applications through 5G technology. Digital Communications and Networks 7(2): 235–244.
45. Cheng K (2020) Smart Rural Financial Innovation based on 5G Network and Internet of Things. Microprocessors and Microsystems: 103500, In Press.
46. Han Y, Park B, Jeong J (2019) A Novel Architecture of Air Pollution Measurement Platform Using 5G and Blockchain for Industrial IoT Applications. Procedia Computer Science 155: 728–733.
47. Tang Y, Dananjayan S, Hou C, Guo Q, Luo S, He Y (2021) A survey on the 5G network and its impact on agriculture: Challenges and opportunities. Computers and Electronics in Agriculture 180: 105895.

Chapter 2
Energy-Efficient Network Routing Protocols for IoT Applications

Deepa Devassy, Immanuel Johnraja Jebadurai, Getzi Jeba Leelipushpam Paulraj, Salaja Silas, and Jebaveerasingh Jebadurai

2.1 Introduction

IoT takes part a crucial part in our routine life [1]. IoT describes a network that consists of various items that are capable of interacting with each other. Every object contains various components such as sensors, actuators, processors, and transceivers to sense and process the data. Large numbers of such devices called nodes are deployed around the world of the Internet [2].

Every device in the network is connected to its nearest base station either directly or indirectly. In a single-hop network, the nodes are right away joined to the base station, whereas in a multi-hop network, the chunks of packets travel through multiple links to land at the base station. The data packets experience various kinds of delay en route. As the packets travel through the nodes, they process the packet and forward it to the destination. This phenomenon is called routing.

This paper provides an insight into various algorithms and protocols used for routing in IoT. Routing is an energy-intensive process in any node. Improper selection of routing algorithms impacts network lifetime. Hence, the selection of suitable routing algorithms plays a vital role to guarantee network performance. A detailed study of various routing protocols has been carried out and their performance is compared with various parameters such as hop count, network scalability, end to end delay, energy consumption, and packet loss rate. The applicability of those protocols in the IoT environment has also been studied and presented in this paper.

The remainder of this chapter is arranged as follows. Section 2.2 details various issues and challenges in IoT. Section 2.3 presents the classification of various

---

D. Devassy (✉)
Karunya Institute of Technology and Sciences, Coimbatore, India

I. J. Jebadurai · G. J. L. Paulraj · S. Silas · J. Jebadurai
Department of CSE, Karunya Institute of Technology and Sciences, Coimbatore, India
e-mail: getzi@karunya.edu

S. Velliangiri et al. (eds.), *Secure Communication for 5G and IoT Networks*, EAI/Springer Innovations in Communication and Computing, https://doi.org/10.1007/978-3-030-79766-9_2

routing protocols. Section 2.4 addresses the analysis of prominent clustering routing protocols. Section 2.5 identifies various research platforms and tools used in the field of networks. Finally, Sect. 2.6 completes the chapter with future research discussions.

2.2 Issues and Challenges in IoT

IoT in its inherent characteristics poses heterogeneous characteristics. These characteristics include context awareness, scalability, energy consumption, low-powered devices, heterogeneity, network connectivity, mobility, security, load balancing, and congestion control. Insights into these parameters are presented in the following subsection.

- Context awareness
- The context refers to the collected information from any location. The factors such as temperature, pressure, and humidity can be measured through various sensors to determine the context [3]. Then the device analyses the contextual facts provincially to elect the pathway that finest fits the prerequisite of a specified exercise. This type of routing protocol faces mass hurdles all along with information exchange, inclusive of lower delay, larger reliability on data transmission, and least power dissipation.
- Scalability
- Usage of embedded systems in IoT extends huge deployment of tiny devices such as sensors and actuators in online applications. The volume of data generated and bandwidth required, grow unboundedly with the increment in the number of these devices. Hence data management and device management are other difficulties in IoT [4].
- Energy Consumption
- The battery power of electronic components in IoT is drained while transporting data. In view of energy constraint, a tradeoff between the quality of information gained and energy consumption by IoT systems is to be retained. Furthermore, the existence of any resource in IoT relies on the availability of energy. The lack of energy prompts the complete habitat under monitoring. Therefore, there is a remarkable longing to lower energy exploitation for the continued lifespan of resources and the well-planned operation of IoT systems [5].
- Low powered devices
- Novel approaches in telecommunications have granted the design of microchips, which can be linked to the Internet, licensing a new era of Internet-of-things (IoT) [6]. These devices run with either a rechargeable or non-rechargeable battery. It is very essential to maintain the power of these devices, otherwise more nodes will go to an inactive stage, which are called dead nodes. Hence it is very essential to maintain the energy of these low-powered devices in IoT.
- Heterogeneity

- IoT networks allow the communication of a variety of devices using different protocols and standards for different applications. All these devices are heterogeneous [7] in hardware design and specification. Data generated from these devices is also heterogeneous in nature [8]. Hence robust methodologies are required to process and use them.
- Network Connectivity
- Quick connectivity of networks and context-aware arithmetic with network resources is the life of IoT [9]. In IoT, billions of devices are assigned with specific addresses, which are connected worldwide through the internet. This connectivity allows devices to exchange information based on some set of standard communication protocols. Maintaining the connectivity of devices is another major challenge in IoT [10].
- Mobility
- Devices or the nodes connected to the IoT network may be static or mobile. Mobility allows the devices to move around different places [11]. This opens another challenge in IoT implementations. When one device moves from one gateway to another, there are chances of interruption in its service. Hence, the system has to adopt mobility management schemes to provide users with QoS under any network condition. Various mobility mechanisms for ad hoc networks are motivated by birds flying in flocks [12].
- Security and Privacy
- This parameter is not fully addressed in IoT by virtue of the open nature of communication in sensor networks. Networks are more vulnerable to various attacks. Thus, this parameter has vital importance in wireless communication [13]. Security based on DTLS and two-way authentication [14], a technique based on elliptic curve cryptography [15] are some of the secure examples adopted in IoT, which are vulnerable to attack. Hence, some mechanisms to be adopted to ensure security to the data, which is being transmitted using Wi-Fi networks.
- Load balancing
- The exponential growth of IoT devices leads to edge computing. The data is being stored into the cloud as well as retrieved from the cloud. Enormous amount of data is being generated during the course of these devices. Load conditions can fluctuate explosively. Thereby a proper load balancing amidst edge servers is required. Under heavy traffic congestion, distribution of load equally to the edge servers is a major concern [16].
- Congestion Control
- Congestion occurs when more data appears in the network. This causes the dropping of packets or delay in the data transmission. Hence appropriate congestion control mechanisms have to be adopted in various layers of the network model for congestion control [17].

2.3 Classification of Routing Protocols in IoT

Protocols work in sensor networks and wireless networks may be adopted for IoT environments with necessary modifications. This section describes different traits of routing protocols existing in IoT. Routing protocols are mainly classified based on four functionalities. They are packet forwarding, route discovery, routing schemes, and routing operations. Figure 2.1 shows the taxonomy of routing protocols in IoT.

2.3.1 Packet Forwarding

Packet forwarding methods are classified into two. They are

- Hop-by-hop routing: Here, the packets take routes available in the forwarding table of all the nodes. Routing table is updated when the packets travel through it. Each node keeps up a routing table with its incoming and outgoing connections. This link information is updated as and when the network topology changes. The data packet contains only the destination information in its header. Based on this, the link with the least cost metric is selected as the next hop. Thus, it guarantees the network bandwidth [18] by properly allocating the resources.
- Source routing: In source routing [19], the entire route information is being held in the header of every packet. The network resources are already allocated prior to the transmission.

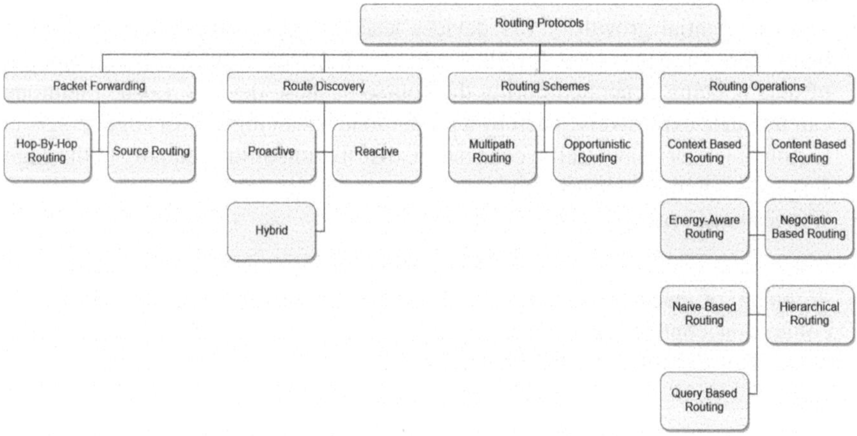

Fig. 2.1 Taxonomy of routing protocols

2.3.2 *Route Discovery*

On the basis of topological information, routing protocols are classified into three. They are

- Proactive: In proactive protocol, nodes know the entire network topology prior to data transmission by sending control packets to their neighbors. The topological information is updated periodically by each node in its routing table.
- Reactive: Lack of awareness of the entire network structure is one of the shortcomings of this protocol [20]. Routes are set up in the network on-demand basis. The routing protocol establishes a new route with other nodes as soon as it joins for communication.
- Hybrid: It is a fusion of both proactive and reactive protocols. The dominant nature of these protocols is exhibited in this routing.

2.3.3 *Routing Schemes*

As per the accessible path from a sender node to the destination machine, the routing scheme can be graded into the following category.

- Single path routing: In this routing, the sender constructs a unique route to the destination [21].
- Multipath routing: In this category, more than one path is available between two nodes. The node can choose any path to reach the destination.
- Opportunistic routing: In opportunistic routing [22], the node immediate to the target is being identified to send the packets.

2.3.4 *Routing Operations*

Context-Based Routing

The context describes the environment of an item. The possible context may be location, connectivity, energy, and so on. It can change vigorously due to the mobility of nodes and fluctuating traffic conditions. This can be another challenge in IoT network. However, the management of context information is restricted by energy or location [23].

Content-Based Routing

The information is forwarded accordant with content [24] or the data generated from the nodes in the IoT network. Various criteria including the existence of fields, specific field values, and so on, define routing. The data is being collected from the network and applies aggregation methods to forward the data.

Energy-Aware Routing

Human interaction has increased in our daily life due to the gigantic climb in low-cost electronic devices. The quality of human life is also raised as a consequence of this. As a result, the lifetime of these devices has been lowered. Energy flawlessness has become the pivotal threat in IoT. Although this main threat has been identified, almost all the solutions have negotiated coverage areas for energy efficiency. Most of the systems demand human interaction for fast recovery from network failures. Human interaction and energy efficiency should be taken care of simultaneously in IoT frameworks [25]. Energy-aware routing protocols lower the energy utilization of nodes in the network. This type of protocol comes under hierarchical routing where the nodes are being arranged in various groups called clusters. The communication reduces the network overhead and improves the lifespan of the network.

Negotiation-Based Routing

The significant concept of negotiation-based routing is the sharing of data obtained from various sensors. It takes spare resources as compared to the meta-data, exactly it is the description of the data sensed by the node. Each node contains resource management to monitor the need of these resources and adapts their functionality accordingly. The best example protocol for this routing is SPIN (Sensor Protocol for Information via Negotiation). It is declared to eliminate defects like flooding and gossiping that arises in other protocols [26].

Naive-Based Routing

Naive routing relies on flooding concepts to find a way to reach the landing place [27]. The sender dispatches a message to all the outgoing links that will flood through all the outgoing links and finally reach the destination. As soon as it arrives at the destination, the acknowledgment is given back to the sender. Even if there is a failure in one link, the request will be flooding through the other links.

Hierarchical Routing

In hierarchical routing [28] the node forms different groups called clusters. The size of the cluster can be fixed or varying. The members will elect a cluster head and the entire communication is through the cluster heads. This type of routing diminishes the energy absorption of nodes.

Analogy of diverse routing protocols is shown in Table 2.1.

The prevailing protocols do not answer the whole concerns and challenges in IoT and well-organized routing is even not attained at all. There is an urge to build novel protocols for conversation, which will satisfy all demands. Effective routing protocols suited to a distinct range of options are needed to find optimal routes [51]. Among these protocols, hierarchical routing protocols use the concept of clustering [52], which will be explained in the next section.

2.4 Analysis of Prominent Clustering Routing Protocols

Clustering algorithms organize nodes of a network into groups or clusters and a specifically designated node, cluster head is devoted for every cluster. Clustering algorithms have been particularly proposed in the context of WSN but their application may also address similar challenges in the Internet of Things (IoT) [53]. Clustering would facilitate energy-efficient routing and topology management by delegating large chunks of communication overhead to cluster heads. There have been clustering algorithms [54] adopted in the research for routing.

Energy consumption of sensor nodes can be minimized by clustering techniques in IoT. All the clustering algorithms are composed to prolong the duration of the sensor network. Consequently, it is very important to address the energy utilization in IoT and the various parameters, which affect the power level of a node. Predominant clustering routing protocols have been analyzed as a separate section with various metrics. The details are shown in Table 2.2.

2.5 Research Platforms and Tools

Various simulation tools or software works on different platforms to scrutinize the execution of these routing protocols. Greater part of the works is simulated by use of NS-2 [61], OPNET [62], Noxim [63], OMNET ++ [64], TOSSIM [65], Monte Carlo [66], HHR [67], and CVX 1.22 [68]. The works have also tested on various platforms such as C++ [67], JAVA [69], MATLAB [70], and OPDMAC [71]. All these hardware platforms and simulators help researchers to choose the best available for a particular application environment.

Table 2.1 Comparison of diverse routing protocols

No table of contents entries found	Classification	Single hop /multi-hop	Proactive/ reactive/ hybrid	Scalability	Mobility	Energy consumption	Delay	Packet loss rate	Application
RPL [29]	Specific protocol for LLN networks	Multi-hop	Proactive	Limited	Limited	Low	High	Low	LLN networks
CoRPL [30]	Specific protocol for LLN networks	Multi-hop	Proactive	Limited	Medium	Low	Low	Low	Cognitive radio networks
E-RPL [31]	Specific protocol for LLN networks	Multi-hop	Proactive	Limited	Limited	High	Low	Low	P2P Communication in LLN networks
mRPL [32]	Specific protocol for LLN networks	Multi-hop	Proactive	High	High	Low	Low	Low	Wireless clinical monitoring applications
DSR [33]	Naive-based routing	Multi-hop	Reactive	Limited	High	Low	High	Low	Mobile ad-hoc networks
DSDV [34]	Naive-based routing	Multi-hop	Proactive	Limited	Low	High	Medium	High	Mobile ad-hoc networks
AODV [35]	Naive-based routing	Multi-hop	Reactive	Limited	High	High	Low	Low	Mobile ad-hoc networks
LEACH [36]	Hierarchical routing	Single-hop	Proactive	Low	Limited	Low	Low	Medium	WSN
FCM [37]	Hierarchical routing	Multi-hop	Proactive	Limited	Limited	Medium	Low	Low	WSN
Cell LEACH [38]	Hierarchical routing	Single-hop	Proactive	Low	Low	High	Low	Low	WSN
ECBR [39]	Hierarchical routing	Single-hop, Multi-hop	Reactive	Limited	High	High	Low	Low	Mobile ad-hoc networks
MINEN [40]	Hierarchical routing	Multi-hop	Proactive	Limited	Limited	Low	Low	Low	IoT, WSN
SPIN [41]	Negotiation-based routing	Single-hop	Proactive	Limited	Limited	Low	Medium	Low	WSN
ESPIN [42]	Negotiation-based routing	Multi-hop	Proactive	Limited	Limited	Low	Low	Low	WSN

No table of contents entries found	Classification	Single hop /multi-hop	Proactive/ reactive/ hybrid	Scalability	Mobility	Energy consumption	Delay	Packet loss rate	Application
Direct Diffusion [43]	Query-based routing	Multi-hop	Reactive	Limited	Limited	Low	Low	Low	WSN
Roumer Routing [44]	Query-based routing	Single-hop	Hybrid	Medium	Limited	Low	Medium	Medium	WSN
ETERNAL [45]	Content-based routing	Multi-hop	Reactive	Limited	Medium	Medium	Low	Low	WSN, IoT
EECBR [46]	Content-based routing	Multi-hop	Reactive	Limited	Limited	Low	Low	Low	WSN, IoT
ELBRP [47]	Energy-aware routing	Multi-hop	Proactive	Limited	High	Low	Low	Low	Mobile ad-hoc networks
FROMS [48]	Energy-aware routing	Multi-hop	Proactive	Limited	Limited	Low	Low	Low	WSN
EHARA [49]	Energy-aware routing	Multi-hop	Proactive	Limited	Limited	Low	Medium	Low	IoT
NLEE [50]	Energy-aware routing	Multi-hop	Proactive	Limited	Limited	Low	Low	Low	IoT

Table 2.2 Analysis of clustering routing protocols

Protocol	Clustering algorithm	Metric for cluster head selection	Hop count between cluster head and base station	Relay node	Advantages of clustering algorithm	Disadvantages of clustering algorithm
LEACH [55]	Randomized clustering	Threshold function	Single hop	Cluster head	Location information of the nodes to create the cluster is not required	Uneven distribution of clusters
Cell LEACH [56]	Randomized clustering	Highest residual energy level	Multi-hop	Cell head and cluster head	Global knowledge of the network is not required	Clusters are distributed randomly
FCM [57]	Fuzzy method	Highest residual energy level	Single and multi-hop	Cluster head	Energy balance among nodes in the cluster	Homogeneous dataset
ECBR [58]	Density-based spatial clustering of applications with noise (DBSCAN) [59]	Highest residual energy level, distance b/w node, and base station	Single and multi-hop	Cluster head	It can discover arbitrarily shaped clusters	Fails to identify the clusters when density is varying
MINEN [60]	K-means, Gaussian Mixture Model (GMM) clustering	Highest residual energy level	Single hop and multi-hop	Cluster head	Large datasets are processed easily	Sensitive to noise

2.6 Summary and Conclusion

Selection of optimal routes is hard in the arena of WSN and IoT. The adaptation criteria depend on numerous parameters. Henceforth, distinct characteristics of routing protocols have been touched on here. Routing is a wide area with so many barriers, which had been discussed thoroughly in this chapter. The review emphasizes that the total network lifetime can be enhanced by forming various clusters in the network. Optimal selection of cluster head is a significant parameter for network routing. In the future, various bioinspired algorithms can be applied to solve the routing issues.

References

1. K. Ashton. 2009. That 'internet of things' thing in the real world, things matter more than ideas, *RFID Journal*, http://www.rfidjournal.com/article/print/4986.
2. J. Sterle, U. Sedlar, M. Rugelj, A. Kos, M. Volk. 2016.Application-driven OAM framework for heterogeneous IoT environments, Int. J. Distribut. Sensor Netw.
3. Harilton da Silva Arajo, Raimir Holanda Filho, Joel J. P. C. Rodrigues, Ricardo de A. L. Rabelo, Natanael de C. Sousa, Jos C. C. L. S. Filho and Jos V. V. Sobral. 2018. A Proposal for IoT Dynamic Routes Selection Based on Contextual Information. *Sensors* 18, 353.
4. P. Misra. Build a Scalable Platform for High-Performance.URL http://www.tcs.com/SiteCollectionDocuments/About%20TCS/Scalability-IoTApplications-0616.pdf.
5. N. Kaur, S.K. Sood.2017. An energy-efficient architecture for the internet of things (IoT), *IEEE Syst. J.* 11: 796–805.
6. Jörg Henkel, Santiago Pagani, Hussam Amrouch, Lars Bauer, Farzad Samie. 2017. Design, Automation & Test in Europe Conference &Exhibition, IEEE *Xplore*.
7. F. Ganz, P. Barnaghi, F. Carrez.2013. Information abstraction for heterogeneous real world internet data, *IEEE Sens. J.* 13 (10): 3793–3805.
8. C. Campolo, A. Iera, A. Molinaro.2014. Named data networking for IoT: an architectural perspective M. Amadeo, in European Conference on Networks and Communications (EuCNC), pp. 1–5.
9. J. Gubbi, R. Buyya, S. Marusic, M. Palaniswami.2013. Internet of Things (IoT): A vision architectural elements and future directions, *Future Gener. Comput. Syst.*, vol. 29, no. 7, pp. 1645–1660.
10. M. Amadeo et al. 2016. Information-centric networking for the Internet of Things: Challenges and opportunities, *IEEE Netw. Mag.*, vol. 30, no. 2, pp. 92–100.
11. Y.-K. Hsiao, Y.-W. Lin.2014. A Mobility management scheme for the internet of things, in Mobile, Ubiquitous, and Intelligent Computing. Lecture Notes in Electrical Engineering, pp. 569–575.
12. S. Misra and P. Agarwal. 2012. Bio-inspired group mobility model for mobile ad hoc networks based on bird-flocking behavior, *Soft Comput.*, vol. 16, no. 3, pp. 437–450.
13. Li, J., Liao, X. & Puech, N.2019. Security and privacy in IoT communication. *Ann. Telecommun.* 74, 373–374.
14. T. Kothmayr, C. Schmitt, W. Hu, M. Brünig, G. Carle.2013. DTLS based security and two-way authentication for the internet of things, *Ad Hoc Networks* 11 (8): 2710–2723.
15. S. Kumari, M. Karuppiah, A.K. Das, X. Li, F. Wu, N. Kumar. 2017. A secure authentication scheme based on elliptic curve cryptography for IoT and cloud servers, *J. Supercomputing*, pp. 1–26.
16. Mogi, R., Nakayama, T., &Asaka, T. 2018. Load Balancing Method for IoT Sensor System Using Multi-access Edge Computing. Sixth International Symposium on Computing and Networking Workshops (CANDARW). https://doi.org/10.1109/candarw.2018.00023
17. A. Betzler, C. Gomez, I. Demirkol, J. Paradells. 2016. CoAP Congestion Control for the Internet of Things.
18. Ronghui Hou, King-Shan Lui, Fred Baker, and Jiandong Li. 2012. Hop-by-Hop Routing in Wireless Mesh Networks with Bandwidth Guarantees, *IEEE Transactions on Mobile Computing*, vol. 11, pp. 264–277.
19. Giovanni Toso, Riccardo Masiero, Paolo Casari, Maksym Komar, Oleksiy Kebkal, Michele Zorzi. 2018. Revisiting Source Routing for Underwater Networking: The SUN Protocol, *IEEE Access* vol. 6, pp. 1525–1541.
20. S. Marwaha, Chen Khong Tham, D. Srinivasan. 2003. Mobile agents based routing protocol for mobile ad hoc networks, Global Telecommunications Conference, 2002. GLOBECOM '02. IEEE .
21. Pravin R Satav, Pradip M. Jawandhiya. 2017. Review on single path multi-path routing protocol in manet: A study, *IEEE XPlore*, pp.5090–2806.

22. Weiwei Chen, Chin-Tau Lea, Shiming He and Zhe Xuan Yuan. 2017. Opportunistic Routing and Scheduling for Wireless Networks, *IEEE Transactions on Wireless Communications*, vol. 16, pp. 320–331.
23. Wenning, Bernd-Ludwig. 2010. Context-Based Routing in Dynamic Networks, Springer ,ISBN: 978-3-8348-1295-7.
24. E. Yoneki, J. Bacon. 2004. Content-based routing with on-demand multicast, 24th International Conference on Distributed Computing Systems Workshops, Proceedings. August 2004, https://doi.org/10.1109/ICDCSW.2004.1284123.
25. Noman Shabbir, Syed Rizwan Hassan. 2017. Routing Protocols for Wireless Sensor Networks (WSNs), *Wireless Sensor Networks*, https://doi.org/10.5772/intechopen.70208.
26. O.U. Akgul, B. Canberk.2016. Self-organized things (SoT): an energy efficient next generation network management, *Comput. Commun.*74 : 52–62.
27. Sriram Sankaran, Ramalingam Sridhar. 2015. Modeling and Analysis of Routing in IoT Networks, IEEE Conference (CoCoNet'15).
28. W.T. Tsai, C.V. Ramamoorthy, W.K. Tsai and O. Nishiguchi. 1989. An adaptive hierarchical routing protocol, IEEE Transactions on Computers, vol. 38, pp. 1059–1075.
29. T. Winter. 2017. RPL: IPv6 Routing Protocol for Low-Power and Lossy Networks, RFC 6550, Accessed on. [Online]. Available: https://rfc-editor.org/rfc/rfc6550.txt.
30. Adnan Aijaz, Hongjia Su, and Abdol-Hamid Aghvami. 2015. CORPL: A Routing Protocol for Cognitive Radio Enabled AMI Networks, *IEEE Transactions on Smart Grid*, vol. 6, pp. 477–485.
31. Abdelhak Zier, Abdelhafid Abouaissa, Pascal Lorenz. 2019. E-RPL: A Routing Protocol for IoT Networks, IEEE *Xplore*.
32. Fotouhi Hossein, Daniel Moreira, Mrio Alves. 2015. mRPL: Boosting mobility in the Internet of Things, *Ad Hoc Networks*, vol. 26.
33. D. B. Johnson and D. A. Maltz. 2004. The Dynamic Source Routing Protocol for Mobile Ad Hoc Networks (DSR), IETF Internet draft, http://www.ietf.org/internet-drafts/draft-ietf-manet-dsr-10.txt.
34. C.E. Perkins & P. Bhagwat. 1994. Highly Dynamic Destination Sequence-Vector Routing (DSDV) for Mobile Computers, *Computer Communication Review*, 24(4), pp. 234–244,
35. C. E. Perkins et al. 2003. Ad hoc On-Demand Distance Vector (AODV) Routing, RFC 3561, http://www.ietf.org/rfc/rfc3561.txt.
36. Asaduzzaman; Hyung Yun Kong. 2010. Energy efficient cooperative LEACH protocol for wireless sensor networks, *IEEE Journal of Communications and Networks*, vol. 12,pp. 358–365.
37. C Hoang, R Kumar, and SK Panda. 2010. "Fuzzy c-means clustering protocol for wireless sensor networks" In Industrial Electronics (ISIE), 2010 IEEE International Symposium on, pages 34773482.
38. Arezoo Yektaparast, Fatemeh-Hoda Nabavi, Adel Sarmast. 2012. An improvement on LEACH protocol (Cell-LEACH), 14th International Conference on Advanced Communication Technology (ICACT), IEEE *Xplore*.
39. Ashish Srivastava, Shraddha Upadhyay, Atul Mishra. 2014. A Conceptual Overview of Energy Consumption based Routing protocol – ECBR. International Journal of Computer Applications 92(3).
40. Vidushi Vashishthz, Anshuman Chhabrayx, Anirudh Khannay, Deepak Kumar Sharma. 2019. An energy efficient routing protocol for wireless Internet-of-Things sensor networks, arXiv:1808.01039v2 [cs.NI].
41. Zeenat Rehena, Sarbani Roy, Nandini Mukherjee. 2011. A Modified SPIN for Wireless Sensor Networks, 978-1-42448953-4/11, IEEE.
42. Jing Li, Chong Shen, 2014. Energy conservative Wireless Sensor Networks for black pepper monitoring in tropical area, 2013 IEEE Global High Tech Congress on Electronics.

43. Velliangiri, S., Sekar, R., & Anbhazhagan, P. (2020). Using MLPA for smart mushroom farm monitoring system based on IoT. International Journal of Networking and Virtual Organisations, 22(4), 334–346.
44. D Braginsky, D. Estrin. 2002. Rumor routing algorithm for sensor networks[C], Proc of the 1st Workshop on Sensor Networks and Applications, pp. 22–31.
45. M. Durga Devi, K. Geetha, K. Saranya Devi. 2017. Content Based Routing Using Information Centric Network For IoT, *Science Direct Procedia Computer Science*, 115, 707714.
46. Chelloug, S.A. 2015. Energy-Efficient Content-Based Routing in Internet of Things, *Journal of Computer and Communications*, 3, 9–2
47. Velliangiri S., Kasaraneni K.K. (2020) Machine Learning and Deep Learning in Cyber Security for IoT. In: Kumar A., Paprzycki M., Gunjan V. (eds) ICDSMLA 2019. Lecture Notes in Electrical Engineering, vol 601. Springer, Singapore. https://doi.org/10.1007/978-981-15-1420-3_107
48. Förster A, Murphy AL. 2011. FROMS: Feedback routing for optimizing multiple sinks in WSN with reinforcement learning. *Ad Hoc Netw*. 9(5): 940–65. https://doi.org/10.1016/j.adhoc.2010.11.006.
49. Thien Duc Nguyen, Jamil Yusuf Khan, and Duy Trong Ngo. 2018. A Distributed Energy-Harvesting-Aware Routing Algorithm for Heterogeneous IoT Networks, IEEE Transactions on Green Communications, https://doi.org/10.1109/Tgcn.2018.2839593
50. Shailja Agnihotri, K.R. Ramkumar. 2017. A survey and comparative analysis of the various routing protocols of Internet of Things, *International Journal of Pervasive Computing and Communications*, ISSN: 1742-7371.
51. Vellanki M, Kandukuri SPR and Razaque. 2016. A Node Level Energy Efficiency Protocol for Internet of Things, *J Theor Comput Sci* 3: 140. https://doi.org/10.4172/2376-130X.1000140.
52. Sahil Sholla, Sukhkirandeep Kaur, Gh Rasool Begh, Roohie Naaz Mir and M Ahsan Chishti. 2017. Clustering Internet of Things: A Review, *Journal of science and technology*: Issue on information and communications technology, vol. 3, No. 2.
53. S. Velliangiri, G. K. L. Kumar and P. Karthikeyan, "Unsupervised Blockchain for Safeguarding Confidential Information in Vehicle Assets Transfer," 2020 6th International Conference on Advanced Computing and Communication Systems (ICACCS), Coimbatore, India, 2020, pp. 44–49, https://doi.org/10.1109/ICACCS48705.2020.9074285.
54. D Wei, H Anthony Chan, 2006. Clustering Ad Hoc Networks: Schemes and Classifications, IEEE.
55. V. K. Arora, V. Sharma, M. Sachdeva. 2016. A survey on LEACH and otherŠs routing protocols in wireless sensor network, *Optik—Int. J. Light Electron Opt.*, vol. 127, no. 16, pp. 6590–6600.
56. R. P. Mahapatra, R. K. Yadav. 2015.Descendant of LEACH based routing protocols in wireless sensor networks, *Proc. Comput. Sci.*, vol. 57, pp. 1005–1014.
57. Saurav Ghosh, Sanjoy Mondal, Utpal Biswas. 2017. Fuzzy C Means based Hierarchical Routing Protocol in WSN with Ant Colony Optimization, 2nd International Conference on Applied and Theoretical Computing and Communication Technology (iCATccT),
58. Cao, F. and Singh, J.P. 2004. Efficient Event Routing in Content-based Publish-Subscribe Service Networks. Proceedings of the Twenty-Third Annual Joint Conference of the IEEE Computer and Communications Societies, Hong Kong, 929–940.
59. Schubert, E., Sander, J., Ester, M., Kriegel, H. P., & Xu, X. 2017. DBSCAN revisited, revisited: why and how you should (still) use DBSCAN. In ACM Transactions on Database Systems (TODS), 42(3), 19.
60. Jun Huang, Yu Meng, Xuehong Gong, Yanbing Liu, and Qiang Duan. 2014. A novel deployment scheme for green internet of things. *IEEE Internet of Things Journal*, 1(2):196–205.
61. T. Meng, F. Wu, Z. Yang, G. Chen, A.V. Vasilakos. 2016. Spatial reusability-aware routing in multi-hop wireless networks, IEEE Trans. Comput. 65 (1), 244–255.
62. Velliangiri, S. (2020). An Enhanced Multimedia Video Surveillance Security Using Wave let Encryption Framework. Journal of Mobile Multimedia, 15 (3), 239–254.

63. H.K. Hsin, E.J. Chang, K.Y. Su, A.Y. Wu. 2015. Ant colony optimization-based adaptive network-on-chip routing framework using network information region, IEEE Trans. Comput. 64 (8), 2119–2131.
64. A. Liu, Z. Zheng, C. Zhang, Z. Chen, X. Shen .2012. Secure and energy-efficient disjoint multipath Routing for WSNs, IEEE Trans. Veh. Technol. 61 (7) 3255–3265.
65. Velliangiri, S., & Karunya, P. K. (2020, January). Blockchain Technology: Challenges and Security issues in Consensus algorithm. In 2020 International Conference on Computer Communication and Informatics (ICCCI) (pp. 1–8). IEEE.
66. Y. Yang, R.S. Blum, B.M. Sadler. 2009. Energy-efficient routing for signal detection in wireless sensor networks, IEEE Trans. Signal Process. 57 (6) 2050–2063.
67. D. Jie, W. Xiong, W. Sheng, X. Shizhong. 2015. HHR: hierarchical hybrid routing scheme for information-centric network, China Commun. 12 (6) 141–153.
68. X. Zhang, Y. Zhao, S. Wang, S. Xu, X. Wang, X. Gao, C. Qiao.2015. Multiobjective optimization for green network routing in game theoretical perspective, IEEE J. Sel. Areas Commun. 33 (12) 2801–2814.
69. R. Dai, P. Wang, I.F. Akyildiz. 2012. Correlation-aware QoS routing with differential coding for wireless video sensor networks, IEEE Trans. Multimedia 14 (5) 1469–1479.
70. R.K. Guha, C.A. Gunter, S. Sarkar. 2007. Fair coalitions of power aware routing in wireless networks, IEEE Trans. Mob. Comput. 6 (2) 206–220.
71. O. Basan, M. Jaseemuddin. 2011. A conflict analysis framework for QoS-aware routing in contention-based wireless mesh networks with beam forming antennas, IEEE Trans. Wireless Commun. 10 (10) 3267–3277.

Chapter 3
Cybersecurity in 5G and IoT Networks

Vani Rajasekar, J. Premalatha, and Muzafer Saracevic

3.1 Introduction to 5G and IoT

Upcoming innovations, such as IoT (Internet of Things), blockchain, cloud computing, and so on, are indeed the outcome of the ongoing rise in WWW bandwidth (i.e., 5G networks) [1], thereby gradually connecting devices to the internet/www. Currently, IP addressing demands concentration to uniquely identify these devices. Thus, it is appropriate to switch from IPV-4 (internet protocol version-4) to IPV-6 (internet protocol version-6) to resolve this growing number of users over the internet (www), and next-generation Internet Protocol (IP) being adequate to discuss more users across the internet. Currently, IP addressing demands concentration to uniquely identify these devices. Thus, it is appropriate to switch from IPV-4 (internet protocol version-4) to IPV-6 (internet protocol version-6) to resolve this growing number of users over the internet (www), and next-generation Internet Protocol (IP) being adequate to discuss more users across the internet. Currently, due to the high number of devices over the Internet and the rising bandwidth [2] of the site, ever more users are being linked to the web, which raises the likelihood of security breaches and attacks from the digital realm.

V. Rajasekar (✉)
Department of Computer Science and Engineering, Kongu Engineering College, Erode, Tamil Nadu, India

J. Premalatha
Department of Information Technology, Kongu Engineering College, Erode, Tamil Nadu, India
e-mail: jprem@kongu.ac.in

M. Saracevic
Department of Computer Sciences, University of Novi Pazar, Novi Pazar, Serbia
e-mail: muzafers@uninp.edu.rs

S. Velliangiri et al. (eds.), *Secure Communication for 5G and IoT Networks*, EAI/Springer Innovations in Communication and Computing,
https://doi.org/10.1007/978-3-030-79766-9_3

3.1.1 Overview of 5G and IoT

Recently, not only are the physical entities acquiring the features from new technology in this IoT period but also from the digital side. This convergence of digital technologies and physical objects leads to partnerships among a wide range of industries. Countless studies are conducted to measure and forecast the material effect of the fifth generation (5G) and the Internet of Things (IoT), some of which concentrate on the value chain others on the benefit to society. And even as such researches are underway, it is clear that 5G would spur innovation in several fields. 5G establishes the basis for realizing IoT's full potential. Although till about 2020, 5G is scheduled for commercial availability, the market works now on the implementation of new global standards and pre 5G products to available anywhere in the industries.

3.2 Requirements of 5G and IoT

The intense trend in 5G-IoT and its information analytics components make people understand precisely its conceptual structure, possibilities, and difficulties. Having taken its particular attributes into account, IoT data presents emerging challenges for both 5G and IoT [3], including trust, protection, and privacy, and even has a direct effect on computational cost and complexity in the storage of data processing facets. In the meantime, the transmission quality and the channel usage are also difficult in different situations. The capacity to process large amounts of data and optimize the communication protocol, therefore, requires smarter methods, such as machine learning techniques. 5G and IoT is an internet-connected network using networking strategies of the next century for data transmission and processing.

3.2.1 Architecture of 5G and IoT

The Internet, Intelligence, and Things (IoT) are all coming together in 5G and IoT. The Internet of Things (IoT) is a concept that includes a large number of interconnected network entities. The convergence of intelligence and information is a smart person, with the goal of building high structural agents or devices to realize difficult requirements including image recognition, natural language processing (NLP), and smart control. Machine learning and deep learning become potentially promising methods for meeting those targets. Centered on certain key technologies, in particular, the 5G communication protocols [4] and levels of data exchange, such as the convergence of the Internet and knowledge, have higher communication capacities on the smart Internet. The study objectives are designed to allow the 5G I-IoT to handle data using artificial intelligence (AI) methods. The object converter not only links sensing domains and processing centers but also manages the relevant information gathered from sensing regions and transfers the same to a distribution center. It introduces numerous innovative technologies, including large MIMO networks, dense static single-cell networks, and small mobile cells networking. The objective of the 5G I-IoT is to create a picture where artifacts are incorporated dramatically into real time and knowledge area to gain various uses, such as smart cities, agriculture, and energy and water conservation (Fig. 3.1).

3.3 Crowdsourcing Use Cases and Evolution to 5G IoT Cybersecurity

In this section, the crowdsourcing uses cases that could be assimilated for use in 5G protection, which will be addressed.

3.3.1 Use Case for Commercial Purpose

Part of the network structure may be availed by multiple service providers (SPs) as a result of the 5G business model. Security measures must be implemented to exchange sensitive information amongst SPs to remove the consequences of potential dissemination of threats while confirming that such services comply with the requirements of privacy and confidentiality and some other ethical considerations.

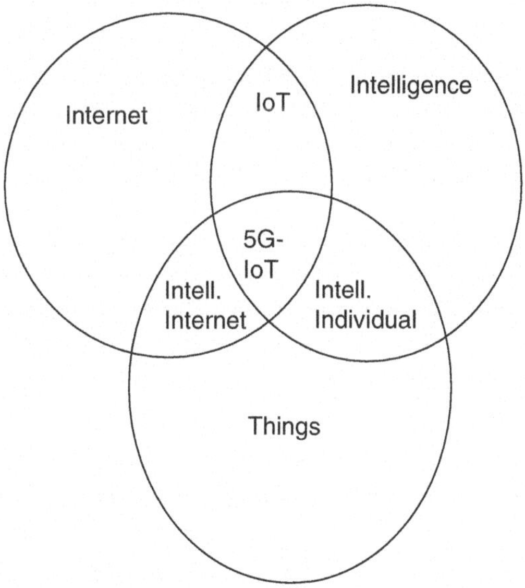

Fig. 3.1 Architecture of 5G IoT

This will be crucial to ensuring that the shared knowledge does not enable corporate surveillance. A common crowdsourcing methodology could be used in frameworks such as the Malware Information and Sharing Platform (MISP). This is a technology platform for exchanging data among organizations on significant risks, encouraging the creation of protective measures by creating virus protection Compromise Indicators, firewalls, and modules for Intrusion Detection.

3.3.2 Mitigation of Physical Attacks

Data are created in crowdsourcing from mobile phones with power-contained agent detectors to provide an early warning system (EWS) within a network security system. The preferred method relies on the architecture of sensor nodes, which send signals for countering physical attacks to a secure network. Software attacks vary significantly from physical attacks, which involve the review of new models. One of the key issues of 5G networks [5] is that the technology is still unable to retain reliable information about IoT devices over a certain period. Consequently, after a while, the controllers neglect if a malicious attacker was an input feature node to the cell. Physical layer protection mechanisms try to solve these problems but rely on network knowledge they actually don't have. Another possibility to address or minimize the issues is to enforce particular security measures at the edge or on the IoT devices of the consumer or to incorporate crowdsourcing frameworks among reliable devices to identify possible threats.

3.3.3 Collaboration Among End Users

Crowdsourcing is important as a platform for end-user partnership toward threats on mobile terminals. The definition of digital witness (DW), for IoT environments, is established in this regard. The goal is to encourage people to track misconduct, threats, and crimes toward individuals or other citizens and to preserve potential data to be used in a court of law. This solution needs a person's participation to stop certain attacks as we cannot see people but our machines can identify them. Only at the edge of a 5G IoT network, this logic can be extended, delivering DW features as a utility. Similar principles can be used to collect electronic proof from vehicle networks (cars as witnesses), which is very useful in view of the fact that vehicle networks represent the main use case in 5G.

3.4 Crowdsourcing for 5G IoT Cybersecurity

Crowdsourcing can be categorized according to different factors that identify eight basic features, namely, (1) for a clearly defined crowd, (2) if there is a task with a clear goal, (3) the incentive the crowd is getting is obvious, (4) the crowdsource is obviously understood, (5) the compensation that the crowdsource is to obtain is explicitly stated, (6) it is an online participatory mechanism allocated to, (7) it provides a reference variable extent call, and (8) the Internet is used whilst the idea has expanded to provide independent standards are still being implemented by organizations (e.g., sensors).

3.4.1 IoT Devices in 5G—A Capillary Network

IoT devices form a capillary network in 5G in which some devices are easy to connect with other nodes via D2D (device to device) communication. Extending this definition to machine-type contact (MTC) in 5G [6]. The systems have multiple opportunities to integrate into the 5G networks. There are various ways of linking devices to the 5G networks: (a) Aggregation, (b) Direct, and (c) Short-range D2D. Ultra-reliable MTC (uMTC) is needed for essential services including road safety and industrial manufacturing. Furthermore, 5G generates greater relays to improve the range of the antennas, while avoiding barriers to signal propagation. Furthermore, the devices or computers must use the relays to attach to the antennas. The capillary network devices may use crowdsourcing methods to alert about D2D abusive behavior and regional threats. The crowdsourcing will become the residual RATs in which IoT devices are eventually orchestrated to access the 5G networks. Notice that it is essential to guarantee the credibility of the machines to make this a reality, and perhaps not all the devices will provide accurate information about the threat. In addition, local security measures on IoT devices would need to be deployed in 5G. There are many reasons for this, but the sheer number of devices makes it nearly impossible to manage and measure all of the congestion in real time at the center. Furthermore, once this traffic passes through the capillary network and enters the heart, it can be dangerous. After this, SDN allows for the application of various requirements to different use cases and flows. This allows congestion optimization and control systems to be applied in particular network segments depending on the evidence generated by the crowds.

3.4.2 Dynamic Ran Enablers in 5G IoT

The IoT modules in the capillary network link to the 5G infrastructure's spatial components (e.g., strong 5G antennas and relays) when using radio channel access technologies (RATs) (e.g., LTE-A, mmWave). One of the changes in 5G would be that it uses DyRAN to allow devices, among several other benefits, to serve as guaranteed access nodes. This aspect of the 5G architecture is especially sensitive, specifically in the context of relay networks, to direct attacks. In 5G, relays are very necessary because they enable the antenna range to be expanded. Furthermore, the contact between both the IoT devices in a particular area and the SPs is disrupted if the relays are disturbed. In addition to physical threats, communication may be disrupted by replacing the ground station nearest to the relay (or any DyRAN device) and dropping traffic [7], or by blasting the DoS attack. Crowdsourcing will help to alleviate most of these security problems unless the DyRAN enablers will preserve and exchange information about the current state of relays. At the center of the node, this data will be stored, since relays, as well as the remainder of the elements, are part of the 5G infrastructure, it is necessary to be notified as soon as possible.

3.5 Entity-Based Crowdsourcing Model

In a model for mitigating cyberattacks via crowdsourcing, the main components to be discussed are shown in Fig. 3.2.

1. Crowd

   Provide the data as per predetermined principles and crowdsourcing agreements. There may be users and devices for the 5G networks, either public or private.
2. Crowdsources

   Ascertain the strategy for crowdsourcing and provide incentive management.
3. Processing Units

   Data processing elements and the provision of answers. Effective intervention units, when assigned to the edge (e.g., by the MEC) or remote detection units, when assigned to the center of the network.
4. Impartial, External Sources

   External trusted groups, agents responsible for offering additional resources that the crowd sources do not include.

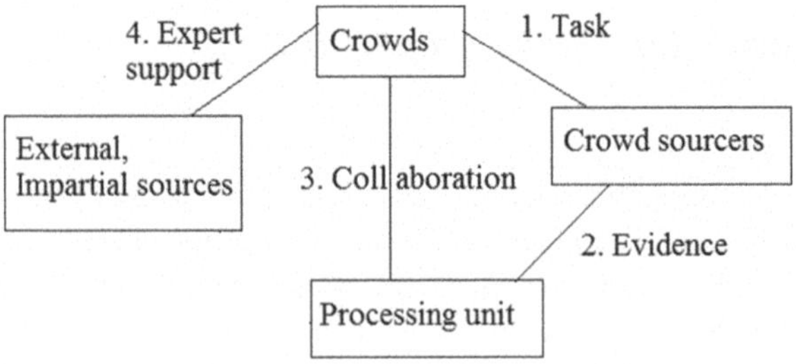

Fig. 3.2 Model for mitigating cyber-attacks via crowd sourcing

3.5.1 *Applying the End-User Collaboration Crowdsourcing Model*

In this scope, it is possible to connect various types of IoT devices (e.g., sensors, cars, cell phones, and drones) to the services offered by the 5G networks (e.g., traffic control information, video, and autonomous information). There are three types of 5G IoT that can be mapped here in this scenario. They are (a) users or service providers, (b) D2D 5G network, and (c) digital witnesses.

Users or Service Providers

The task and compensation are declared by the service provider (SP) and a potential service for those users who want to cooperate in the experiment of crowdsourcing. Traditional visual proof (e.g., pictures of vandalism) is submitted by the user. This evidence is obtained manually, for instance, by using the device's camera. The SP can request the LEA's involvement, depending on the type of evidence and the importance, or possibly use this approach to develop other specific measures to improve the 5G infrastructure.

D2D Devices for 5G Infrastructure

In this situation, the job is to warn about wrongdoing and network attacks. Communication is carried out at a low level and does not enable the consumer to participate. The IoT device's job is to provide details about local attacks and vulnerabilities in the devices' applications. The supervisor of this RAT function tests the data with local data. It contacts the SP if it requires more information or if more analysis is needed of the data supplied by the devices.

Digital Witness—LEA's

As a function, the digital witnessing (DW) feature is supported. The goal is to suggest crowdsourcing platforms that are capable of alerting other users in a particular area of a potential attack. DW produces steam proof validated using the security measures authorized by the LEA (electronically stored information in the computer, such as logs or data about their requirements, and can include personal data). The LEA analyses the data and takes countermeasures that may cause legal proceedings to be initiated. The LEA can contact the SP if necessary to obtain more information or collaborate to avoid such threats to the 5G ecosystem. At this point, certain criteria are apparent. First, it is important to guarantee and preserve the privacy of the people involved in such activities from the rest of the participants in the crowd. Neither would the rest of the audience be public with crucial knowledge about the sources. Finally, vulnerability information should be exchanged only with the users/devices affected.

3.6 Business Model and Cybersecurity

Within the uncertainties of the current business setting, the business model relates to taking abstract objectives to the realistic level of decisions and actions. For several technical companies, business models are the basis, and a well-designed business model will lead to the improvement of even mediocre innovation in the markets but a well-designed technology could fail due to a weak company. The business model has been described specifically as an interconnected architecture set of symbols and an interrelated collection of core logic and strategic decision variables, explaining the content of transactions, transaction management, and mechanisms of transaction relationships for maximized value creation and value capture. Understanding the organizational model is much more relevant for cybersecurity firms, where business failure can lead to serious risks further than just economic risks. On the basis of the study of the trend, it is realized that the potential business effects of cybersecurity in 5G/IoT can be brought to light, which can increase knowledge of how and why business model analysis matters and how. With all of the high energy-connectivity and high-tech technologies and facilities, we are now moving toward 5G, we are constantly exposed to severe cyberattack threats.

Depending on the trend study, it is realized that the potential market impacts of cybersecurity in 5G IoT can be brought to light, which can raise understanding of how and why business model thought matters. With all the hyper-connectivity and high-tech technologies and facilities, we are constantly exposed to serious cyberattack risks, as we are already moving toward 5G. It can be claimed that the more commercialized the protection is, the more businesses are interested in offering safe services and goods by investing more in it. There are basically two types of organizations, from this point of view, whose business depends on security. First, organizations that are interested explicitly or implicitly in the field of security solutions. These businesses market security either as a commodity or as a package with a few other services, products, or facilities. Second, all the other market organizations have impressive digital footprints and face future cyberattack risks. In an imperfect environment in which we operate, any cyberattack would have a direct or indirect effect on one or both of these kinds of business organizations.

3.6.1 Need for Cybersecurity in 5G IoT

Cybersecurity is typically provided as a service or as a product by specific third-party providers or as a hybrid of service and product that helps organizations and people secure their digital content. Companies offering various products/services related to cybersecurity have two wide customer categories. Organization groups, which are collectively referred to as B2B (Business to Business) clients or B2G (Business to Government) clients in marketing literature, have been the more popular customer community so far. In recent times, this community of consumers has shown a rising interest in investing in cyber protection for their business/sustainable development. There is now a steady market potential in this segment because business entities or government agencies are more careful about the protection of their digital properties. The second major customer group, although, is new consumers, known as B2C (business to customer) customers. Up till now, entering this customer community has been relatively difficult for security firms [8–10]. This issue is activated because the importance of the privacy data/information of individual people online and other channels may appear to be expendable by persons even to this day. Although this scenario is more likely to shift in the 5G age, where trillions of Internet of Things. For many organizations, this technique for investing in cybersecurity products/services has been effective, but failing to estimate the need for protection has resulted in fatality many times too often. (IoT) applications are going to be flooded the home setting. The second categories of businesses that are impacted by cybersecurity [11] are those that use cyberspace as an essential feature for their businesses. In the modern world, this category can include almost all kinds of cyber security organizations. Each organizations are dealt to different levels of cyber threats.

3.7 Uses and Benefits of 5G IoT in Smart Grid

The next step in the development of our existing grid networks is the IoT. From D2d communication, IoT communications have developed. IoT aims to use IP-based solutions to link objects on a large scale while enabling them to communicate through the Internet with any other communication device.

3.7.1 Uses of 5G IoT in Smart Grid

The National Institute of Standards and Technology (NIST) describes the smart grid as the convergence of the electric grid from the last generation with the growth of information and communications technology in the present century. Unlike the conventional electricity grid, the smart grid maximizes electricity delivery, improves reliability, reduces losses, and enables large-scale renewable energy sources such as solar and wind farms. The current grid system faces significant challenges, namely frequent dark-outs, peak time overburdening, and service failures that are never recorded in time, largely due to an aging infrastructure [12]. However, the implementation of remotely sensed devices designed to measure, tracking and providing data about the network particular set it more interconnected and knowledgeable. The generated resource data could then be used to implement a self-healing grid, increasing the degree of self-monitoring and decision-making while also enhancing overall efficiency. The smart grid (SG) is the world's largest relatively close IoT network. It is an interactive power grid. From the construction of power stations to the production of energy, the entire power grid chain is involved (houses, buildings, factories, public lighting, electric cars, smart appliances, etc.) [13], to securely monitor and control the power grid, systems such as the transmission of electricity will be fitted with intelligent and two-way communications infrastructure. This will be allowed using smart meters, smart appliances, and sensor systems. The goal of the SG is to maintain a real-time equilibrium among production and demand, by allowing the power chain to be monitored and regulated, thanks to the large number of smart objects interacting in two directions. The smart grid (SG) is deemed one among the most infrastructural developments but is probably the main possible implementation of IoT networks. Countless wireless sensors, smart meters, smart cities, sensors, and other smart objects are incorporated into the configuration of smart grid networks, all of which interact with each other through a connected network. The IoT enables remote sensing and monitoring of devices on the smart grid infrastructure through a scalable communication network.

3.7.2 Advantages of 5G IoT in Smart Grid

Advanced Metering Infrastructure (AMI)

The sophisticated noise reduction infrastructure can be readily implemented using an IoT smart grid. The AMI is in charge of tracking, processing, retrieving, and delivering the meter-in data submitted by smart meters to the utility company's billing, failure management, and demand forecasting servers. The availability of real-time pricing offers valuable indicators for both customers and suppliers, allowing them to better manage their energy needs and resources.

Improved Reliability of Power System

In the event of any overt or implied disturbances or threats, an IoT smart grid will easily recover itself (self-healing). It also supports the network's personal reconstruction, after threats, natural catastrophes, outages, or failures of networking devices, via routing algorithm to restore electricity, the capacity to produce microgrids and autonomously managed land in the case of power failure, and the detection of energy deficiency sources.

SMADA (Supervisory Monitoring and Data Acquisition) Improved Functions

Because of the use of a multitude of sensors, smart meters, and actuators to track the entire system of the power grid, they will report information to both the utility regularly, on demand, or in reaction to certain incidents, while still responding to utility demands owing to their two-way communication capability.

3.8 Cybersecurity Challenges in 5G IoT

There is no question that cybersecurity is indeed one of the greatest and most difficult problems facing 5G IoT networks. Sensors, computers, and networks linked to the Internet are frequent targets of cyber investigation, sabotage, extortion, fraud, and even damage. Therefore, a cyberattack [14] will lead to disastrous consequences and also major financial losses because such an attack would put an end to entire countries.

3.8.1 Classification of Recent Attacks

Cyberattacks can disrupt the physical infrastructure of a utility, make them unusable, cede control of that infrastructure to an external agency, or put the privacy of employees and customer information in jeopardy.

Device Attack

The purpose of a system attack is to compromise and monitor a network-working computer on the grid. It is also the first phase of a major attack in which one compromised system is used as an entry point to initiate additional attacks and damage the entire network of the smart grid.

Data Attack

To deceive the smart grid to make incorrect decisions/actions, a data attack tries to unlawfully insert, change or erase data, or control commands in the communications internet traffic.

Privacy Attack

A privacy attack needs to find about the private or personal data of a user by providing data from the tools of their smart grid network. Such data could include data on energy consumption where low to no energy demand for some periods may be used to conclude that the location is most likely not occupied.

Network Attack

In the type of denial of service (DoS), a network accessibility attack primarily takes place. Its aims to use or overuse the smart grid network's communications and computing cost, culminating in malfunction or interruption in communications [15]. Consequently, a bandwidth allocation attack must be efficiently dealt with. A network attack on an IoT-based 5G could make millions of devices unavailable, making it inoperable as the devices would be available.

3.8.2 *Classification of Recent Attacks*

Potential threats, many of which were completely unseen, are developing every day, making it extremely difficult to eradicate cybersecurity attacks. This makes contingency measures for cyber threats of paramount significance as they include the first line of defense in the event of potential attacks.

Access Control

There are rules that are predefined to determine the services, data files, or devices with which permission should be given to a user or computer and to which access should not be available at all.

Encryption

When implementing 5G IoT solutions, traffic flows through IoT devices and control centers must be encrypted, including the servers of the service provider [16]. It is important to ensure that communications are encrypted using powerful encryption tools as it reduces the capacity of an intruder to either hijack communications or produce valid data to fool the system.

Authentication

This includes the identification of connected devices and the authorization of what each device can do on the network. Improve the application is typically the first stage of a data communication session, often resulting in a shared session key to encrypt and authenticate corresponding data packets and ensure data integrity.

Intrusion Detection System (IDS)

To determine and separate vulnerable networks and/or devices, intrusion detection techniques are therefore essential. This will help to cause monitoring systems to take preventive measures to take necessary steps in advance to prevent any potential attacks or to undo any disruptive changes to the system.

3.9 Major Challenges Hindering the Growth of 5G and IoT Networks

A complete description of the challenges is needed to tap into the full potential of 5G IoT-enabled intelligent grids.

3.9.1 Standardization of 5G IoT Protocols

Although some 5G IoT devices can interact with others, there is still no standardized IoT language. The lack of responsibility has resulted in a fragmented array of solutions that are mashed together instead of an easily organized and consistent set of solutions. Manufacturers of devices will have to choose between different frameworks, and consumers must find out on their own if other devices they want are consistent with it because they are. While there have been efforts at standardization, the construction companies with diverse products remain divided on how to standardize. There are two main groups of standards, one called the Consortium for Open Interconnect, while the other is known as the AllSeen Alliance.

3.9.2 Cognitive Capability in 5G IoT

The increase in the number of related objects translates into more knowledge that needs processing. With the number of connected devices exceeding the human population in 2017 and growing to over 50 billion devices by 2020, there is a question as to how to handle this amount of new data [17]. The amount of data generated by this mammoth smart grid system based on IoT will be enormous. Specific data and analytics software capabilities would be needed to move, store, and analyze such huge volumes of data.

3.9.3 High Speed 5G Networks

In a power system, the 5G IoT implementation includes two-way real-time contact between the different artifacts in a smart grid. Thanks to its scalability and low implementation costs, wireless networking has proven to be the best medium for communication. Interconnection is thus the foundation of 5G IoT networks, and low latency, large bandwidth networks are crucial for the proliferation of IoT networks, with a high range of devices estimated to be up to 50 billion devices. Seamless network connection is needed with essential IoT applications such as mobile cities requiring higher data transfer capabilities and low latencies. IoT growth is also related to the creation and adoption of high-speed 5G networks.

3.9.4 Privacy Issues

The 5G IoT poses special challenges to key components of privacy and data security. Most of these problems arise from incorporating IoT devices into our applications, thus breaching the privileges of other users. This is more apparent in private devices, such as smartphones and smart vehicles with tracking smartphones and electronic toys, smart TVs, and cameras with voice and facial contact characteristics. There are indeed cases of overlapping social and cultural norms and values between IoT system implementations and information gathering activities. The possibility that IoT spam's a wide and international reach of the network makes it difficult for the IoT to establish a global data privacy act.

3.10 IoT Based Security Management System

Embedded computing devices are built into a network of honesty through IoT. IoT cybersecurity is a tool for strategic enhancement to increase the security of the entire environment, which includes all of the improvements involved in IoT.

3.10.1 Cybersecurity Oriented IoT Architecture

IoT security framework can be majorly categorized as (1) Basic three-layer architecture, (2) derived four-layer architecture, and (3) detailed five-layer architecture. Those layers are accessing layer, perception layer, middleware layer, network layer, interface layer, and application layer. Architecture and design frameworks, wireless networking and connectivity, concepts and functionality, interconnected and standardized devices, authentication, lightweight technologies, and so on, are associated with the architecture of IoT-based cybersecurity. The development of the infrastructure includes usability, integrity, availability, scalability, confidentiality, and interoperability of heterogeneous connected devices from a technological point of view. Because cybersecurity in a dependent way can alter or may involve real-time interaction, an adaptive architecture is required to assist devices that communicate dynamically with other IoT stuff.

3.10.2 Cybersecurity Attacks on IoT

The IoT is a global system through which smart devices such as Radio Frequency Identification (RFID) tags and readers, sensors, actuators, smartphones are connected together. Due to the inadequate storage space, power consumption, and computing capabilities, IoT-related products are vulnerable to denial of service (DoS) attacks at each level. DoS attack is a malicious attempt to block access by end-users to the Internet of Things services. Spoofing, modifying, or replaying the identity details of smart devices on the IoT network produces a Replay attack. A time attack is an intruder that steals the time-related encryption key and other significant data.

3.10.3 Attack Taxonomy

The attacks tend to be malicious because of the variety of smart devices, communication protocols, software, and services. High-end and low-end computer attacks are attacks focused on devices. Location-based attacks are attacks that are internal and external. Active and passive attacks are the types, which are based on access level. Intrusion, snooping, alteration, production, replay, and man-in-the-middle attacks involve attacks based on information damage. Host-based attacks are attacks on devices, infrastructure, and applications. Strategy-based attacks are physical and logical attacks. Protocol-based attacks are attempts on interruption and divergence. Interpretation, platform, web services, and device attacks are layer-based. High-end device attacks include high-power/full-fledged devices to initiate IoT network attacks, whereas low-end device threats include low-power devices to target the IoT system. External attackers may claim to be outsiders but might perform on the IoT network malicious software.

3.11 Key Applications of 5G and IoT in Industries

Over the next few years, 5G networks and IoT technologies are designed to be widely deployed. 5G would see rapid implementations in city centers at the realistic stage. This is maybe fortuitous from an IoT point of view, like smart cities, smart universities, and creative hub would be sponsored by many conventional IoT applications.

3.11.1 Smart Grid Applications

An intelligent system is constructed to transmit electricity created effectively although with the required energy consumption efficiency. Smart grid technologies monitor the operation of both power sources and electric grids efficiently. Consumer use or policies are often controlled by smart grid systems and tracked. Therefore, these systems require ultrareliable low-latency communications networks (e.g., the suggested end-to-end bandwidth of the linear energy production pre-phase is 1 ms) to transfer information on the web. By achieving an end-to-end bandwidth of 1 ms with super-high consistency, the Tactile Internet will protect such systems.

3.11.2 Tactile Robots

Various emerging developments in robotics, cross-scale teleportation, wearable devices, distributed computing, and digital applications draw on this new frontier of collaborative coexistence among humans and robots.

3.11.3 Heterogeneous Networks (HtNet)

Heterogeneous Networks is a wide deployment architecture designed to meet the client-driving 5G IoT on-demand specifications. HtNet helps 5G-IoT continue providing requests for on-demand data transfer speeds. 5G-IoT can deploy a wide range of devices that are limited by capital. A variety of Het-Net solutions are being proposed to ensure the Quality of Service (QoS) for 5G-IoT systems: a major MIMO-based Htnet has indeed been built to create large-scale hummers MIMO link. The HtNet was originally intended for human-to-human (H2H) interaction, while Machine-Type Communication (MTC)s implementations are distinguished by a variety of innovative features, including a massive number of sensors, high-frequency network connectivity, and strict QoS specifications during emergencies.

3.11.4 5G and IoT Applications in Smart Cities Deployment

The new Quality of Experience (QoE) had driven 5G IoT T (tactile Internet architecture) for smart cities, comprising five layers: device layer, sensing layer, processing layer, transmission layer, and application layer has been developed to fully integrate tools to control with smart cities. Especially, from each framework of this architectural design, the techniques defined follow the requirements of low latency, high reliability, and high user knowledge. In this framework, a rapid and effective QoE management platform is supposed to focus on a broad development program.

References

1. Wang, Dan, Dong Chen, Bin Song, Nadra Guizani, Xiaoyan Yu, and Xiaojiang Du. "From IoT to 5G I-IoT: The next generation IoT-based intelligent algorithms and 5G technologies." *IEEE Communications Magazine* 56, no. 10 (2018): 114–120.
2. Khurpade, Jyoti Mante, Devakanta Rao, and Parth D. Sanghavi. "A Survey on IOT and 5G Network." In *2018 International Conference on Smart City and Emerging Technology (ICSCET)*, pp. 1–3. IEEE, 2018.
3. Rajasekar, Vani, K. Sathya, and J. Premalatha. "Energy Efficient Cluster Formation in Wireless Sensor Networks Based on Multi Objective Bat Algorithm." In 2018 International Conference on Intelligent Computing and Communication for Smart World (I2C2SW), pp. 116–120. IEEE, 2018.
4. Liu, Xin, and Xueyan Zhang. "Rate and energy efficiency improvements for 5G-based IoT with simultaneous transfer." *IEEE Internet of Things Journal* 6, no. 4 (2018): 5971–5980.
5. Mistry, Ishan, Sudeep Tanwar, Sudhanshu Tyagi, and Neeraj Kumar. "Blockchain for 5G-enabled IoT for industrial automation: A systematic review, solutions, and challenges." *Mechanical Systems and Signal Processing* 135 (2020): 106382.
6. Rajasekar, Vani, S. Varadhaganapathy, K. Sathya, and J. Premalatha. "An efficient lightweight cryptographic scheme of signcryption based on hyperelliptic curve." In *2016 3rd International Conference on Recent Advances in Information Technology (RAIT)*, pp. 394–397. IEEE, 2016.
7. Wei, Cui. "5G-oriented IoT coverage enhancement and physical education resource management." *Microprocessors and Microsystems* (2020): 103366.
8. Rajasekar, V., Premalatha, J. & Sathya, K. Cancelable Iris template for secure authentication based on random projection and double random phase encoding. Peer-to-Peer Netw. Appl. 14, 747–762 (2021). https://doi.org/10.1007/s12083-020-01046-6
9. Rajasekar, Vani, J. Premalatha, and K. Sathya. "Enhanced Biometric Recognition for Secure Authentication Using Iris Preprocessing and Hyperelliptic Curve Cryptography." *Wireless Communications and Mobile Computing* 2020 (2020).
10. Rajasekar, Vani, J. Premalatha, and K. Sathya. "Multi-factor signcryption scheme for secure authentication using hyper elliptic curve cryptography and bio-hash function." *Bulletin of the Polish Academy of Sciences. Technical Sciences* 68, no. 4 (2020).
11. Mavromoustakis, Constandinos X., George Mastorakis, and Jordi Mongay Batalla, eds. *Internet of Things (IoT) in 5G mobile technologies*. Vol. 8. Springer, 2016.
12. Rajasekar, Vani, Premalatha Jayapaul, and Sathya Krishnamoorthi. "Cryptanalysis and enhancement of multi factor remote user authentication scheme based on signcryption." *Advances in Mathematics of Communications* (2020).

13. Rajasekar, Vani, J. Premalatha, K. Sathya, S. Dhanesh Raakul, and Muzafer Saracevic. “An Enhanced Anti-phishing Scheme to Detect Phishing Website.” In *IOP Conference Series: Materials Science and Engineering*, vol. 1055, no. 1, p. 012077. IOP Publishing, 2021.
14. Premalatha, J., and Vani Rajasekar. “9 Industrial Internet.” *Internet of Things: Integration and Security Challenges* (2020): 135.
15. Saračević, Muzafer, Aybeyan Selimi, and Šemsudin Plojović. “Some specific examples of attacks on information systems and smart cities applications.” In *Cybersecurity and Secure Information Systems*, pp. 205–226. Springer, Cham, 2019.
16. Saračević, Muzafer H., Saša Z. Adamović, Vladislav A. Miškovic, Mohamed Elhoseny, Nemanja D. Maček, Mahmoud Mohamed Selim, and K. Shankar. “Data Encryption for Internet of Things Applications Based on Catalan Objects and Two Combinatorial Structures.” *IEEE Transactions on Reliability* (2020).
17. Saračević, Muzafer, Saša Adamović, Nemanja Maček, Mohamed Elhoseny, and Shahenda Sarhan. “Cryptographic keys exchange model for smart city applications.” *IET Intelligent Transport Systems* 14, no. 11 (2020): 1456–1464.

Chapter 4
5G and IoT Networks Risk Management

M. Umaselvi, E. Menaka, V. Chandrasekar, and D. Saravanapriya

4.1 Introduction

5G and IoT is one of the incredible innovations. The popularity of smart technology has gained lots of potential benefits it guarantees. The view of incorporating the use of smart devices in your home can create your life somewhat more accessible, more relaxed, and highly suitable. However, despite all the potential rewards that it brings on board, 5G and IoT also have their fair share of drawbacks. The technology learning curve, inactive way of life, setting up costs, and internet reliance are notable demerits. Lively situations are characterized by changes in their structural design, data flows, and operational procedures. Such modifications can come into view occasionally, constantly, or be event generated. The 5G and IoT risk management are instantiations across various ecosystems, such as risk management for security risk administration, the healthcare environment, and cyber risk, smart home, and smart cities are shown in Fig. 4.1.

M. Umaselvi (✉) · D. Saravanapriya
P.A. College of Engineering and Technology, Pollachi, India

E. Menaka
Vivekanandha College of Engineering for Women, Namakkal, India

V. Chandrasekar
Malla Reddy College of Engineering and Technology, Hyderabad, Telangana, India

S. Velliangiri et al. (eds.), *Secure Communication for 5G and IoT Networks*, EAI/Springer Innovations in Communication and Computing,
https://doi.org/10.1007/978-3-030-79766-9_4

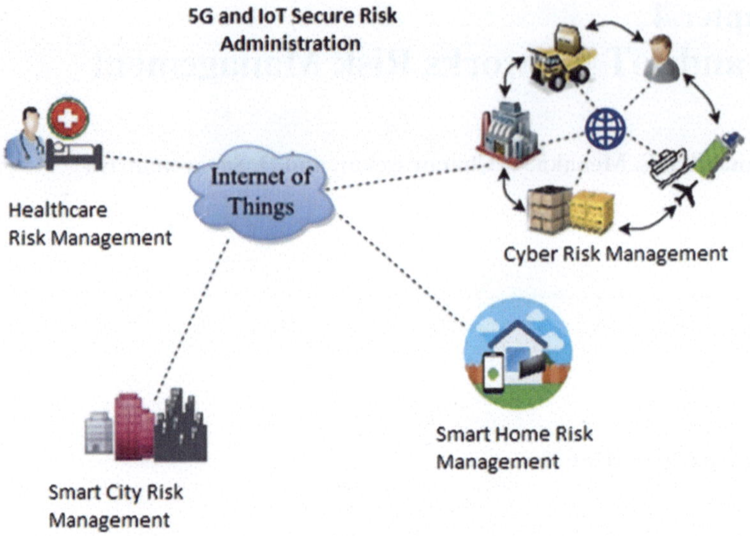

Fig. 4.1 5G and IoT risk administration

4.2 5G and IoT Risk Administration

In like way, risk association can be portrayed as a fragment of the secured structures' advancement strategy, which plan is to see what assets should be restricted, from which hazards, and how these threats could be scattered while bundling revealed perils of the arrangement [1]. Its acting as a most enormous circumstance is growing sensible courses of action subject to the conditions and conceivable security hacktivist. Ordinarily, it offers a lot of rules to lessen the peril level or thoroughly stop conceivable attacks on the structure near, with the suggestion for beneficial and emotional system reviewing. Special methodologies are essential which relate danger association in a collection of regions, that is, the explanation of a significant kind of the system's design is basic during the security risk astute cycle.

4.2.1 5G and IoT Security Controls

All through the peril, the chief strategy in IoT formative likely threats is not the primary development in a safe system. It is a never-ending basic to perceive how to cover reachable threats to moderate conceivable perils. In the IoT, ensuring about methodology needs a sweeping enduring of security controls for all resources of the system. In this segment, we assign and depict land enters in the creation of secure with IoT structures and intend sensible solutions for wrapping one of a kind of systems risks [2].

Network Security for 5G and IoT

Association security frequently required upgrades in cryptography methodologies and suitable key organization plans, as it acts as a most huge capacity in rising security for ensuring about association system. Yet, with appropriate standard peril, the board to the valuable resources of the IoT structure could be inadequate.

Protected System Services

Simply basic ports that are fundamental for the sensible system handiness should be revealed and open. Repair should be ensured about from spread out and lightening assaults and make sure that the threat the board is facilitated and not peril to refusal of organization attacks.

Transport Risk Procedure

Secure Sockets Layer (SSL) and Transport Layer Security (TLS) when delivering information beginning to end association to be certain that your information is mixed. If SSL or TLS is not obtained, do not stop for one moment to use other standard encryption methods to make sure about data. Give a thought about the use of encryption structures. Abstain from using unaccepted encryption regards while managing IoT risk examination frameworks.

Information and Solitude

Information and confinement key as regularly as conceivable speak to the larger part responsive and important resources of the IoT risk board. Every customer needs their individual data to be bound and dealt with without assistance from any other person; anyways, there are a lot of plans that maintain on running customer information. In such belongings, an immense concern is to confine and check the summary of standard information bosses. It should be set up as an explicit running course of action depending on the data type managed by the arrangement.

Solitude Anxiety

Make sure that private data are precisely ensured during all periods of the structure. Provide admission to the made information only for affirmed clients. Do not store up extra information that requires fitting patterns of the affiliation. At all events revivify information before review.

Protect Firmware for IoT Devices

Make sure that the structure utilizes the modernized system of transport depending on the cryptographic techniques. The renewed organizer must not picture responsive information. Every set of obtained points of reference with energies should be stepped and attested prior to saving in system memory additional room. Secure Boot is the place where the structure firmware watches that the system boot loader is supported with a cryptographic key affirmed by a database contained in the firmware.

Physical Safety

Employ just certainty and ensure information procedures that cannot be effortlessly isolated. All responsive plans' information should be mixed at unwinding. For immaculate assurance, connectors or other outside ports are ensured about from moving contemptuous programming. It is prescribed to decrease the proportion of USB ports by the structure. Ensuring the thing can limit administrative limits.

Individuality Administration

Solitary association spoke to a procedure for affirming a stunning assortment of uniqueness and alliance types. We can distribute the main structure which must be searched after through IoT network uniqueness oversight. The uniqueness of thing should perpetually be sole separation to various articles from its relations. Identity can be named focus character as an article can too have more than two or three vaporous characters.

4.3 Healthcare Environment Risk Management

The making of IoT and 5G developments for joint and interrelated clinical equipments and sensors has distorted the circumstance in the clinical administration industry. In any case, with the humility of the dispersed air and clinical devices, IoT and 5G will be the truth of a fundamental where aggressors can see shortcomings and dynamically dispatch their attacks. This ends up being high risk to the clinical administrations' atmosphere which may reason a significant effect on its well-being exertion. Regardless, the public authority assistants of IoT plan in clinical administrations are undisputable. To address this issue, this learning suggests a 5G and IoT security peril to the board model to ensure preparation in the clinical administration atmosphere.

4.3.1 IoT Healthcare Components and Its Risks

5G and IoT unrests in clinical consideration are getting more flexible and sensible in cost. This contains recognizing verification development, correspondence, and domain progresses, distinguishing advances and organization arranged designing [3]. Figure 4.2 depicts how IoT can be applied to the clinical consideration system by using distant sensor association. The chart shows how IoT cloud is executing as a phase in setting up all social events, which are essentially the patient masterminded with all sensor center points, clinical administrations practiced, the patient families, and the emergency office.

The trust of clinical administrations on IoT is generally to improve induction to mind, raise the idea of the thought, and decrease the cost of concern [4]. Despite the rise in clinical administrations' applications reliant on IoT, the delight around these applications far outflanks the reality. Besides, there is a peril that an inordinate measure of using IoT headways will disassociate watchmen from patients, perhaps reason in insufficiency of wise [5]. These obstacles made congruity onto the foreseen security game plans are less reachable. Thus, following subsections will inspect 5G and IoT fragments and their possible risk.

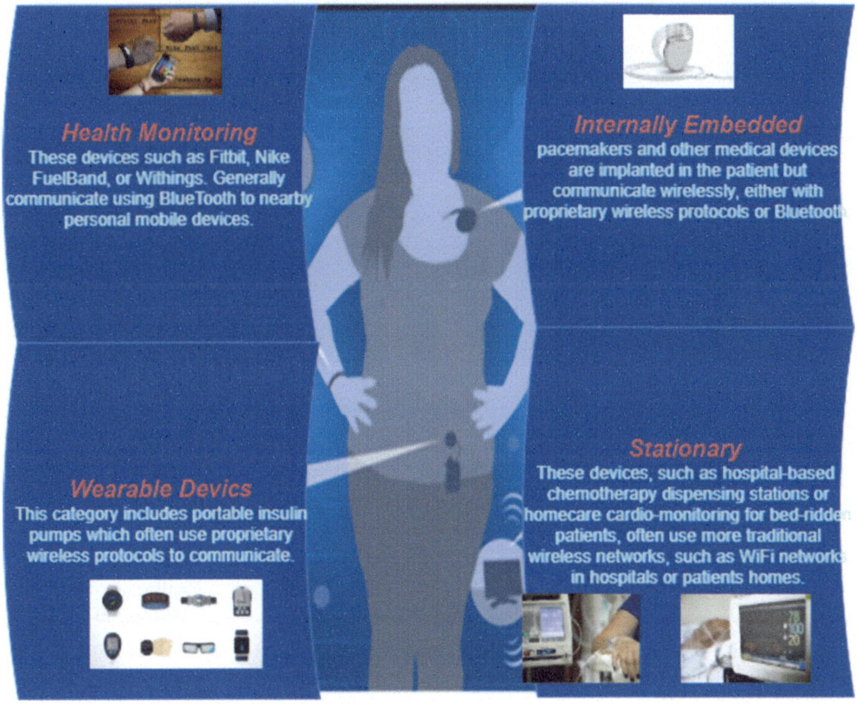

Fig. 4.2 Categories of networked healthcare devices

4.3.2 Clinical Technology Risk Factors

The clinical field is developing quickly. There is moderate growth in interconnected devices nowadays. Generally, for most of the clinical thought, suppliers reviewed in a PricewaterhouseCoopers review held they have included customer fitness like wearable prosperity mindful devices, for instance, motorized medication store association structures with their IT ecology [6]. Despite the way that the basic technique in a huge load of devices midway cover, as Fig. 4.3 shows, network health care devices. The contraptions generally speaking drop into four critical gatherings: customer stock for prosperity seeing, usable outer clinical plans, embedded clinical administration things, and topological clinical consideration devices.

These procedures handle the commitment to convey likewise near and dear and public remunerations in three sorts: (1) It can improve results and significance of life, (2) authorize patients, and (3) cut extended government medical coverage costs. Topological clinical administrations' parts are no different in demonstrating prosperity opening in the breaker of arranged associations (e.g., clinical consideration contraptions), customer geology (e.g., 5G cellulars), and topological date systems (e.g., clinical devices IoT partners). IoT clinical administrations contraptions increase four huge and overlie zones of concern: unconstrained failures, detachment encroachment, orchestrated trouble, and ordinary aggravation.

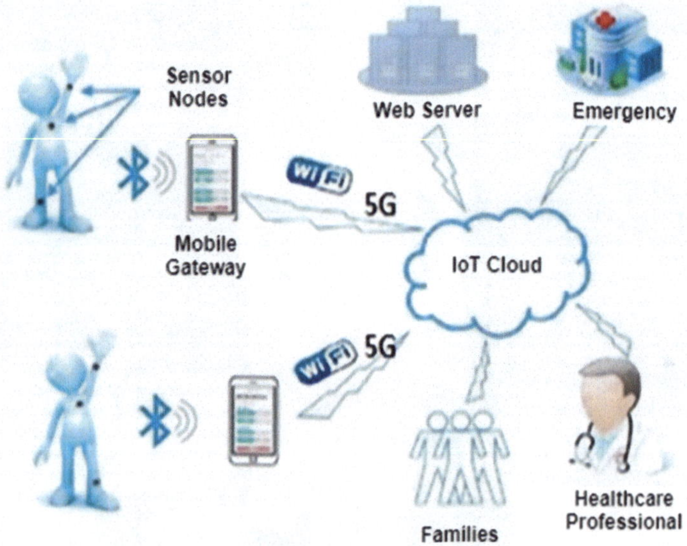

Fig. 4.3 5G with IoT clinical administrations stage

Unconstrained Failures

5G and IoT topological clinical consideration things are peril to more than now unlawful target. Like any further way of thinking, they level to collapse. This is as legitimate for pacemakers. Since the possible deadly issues of clinical administrations contraptions damaged lead dissatisfaction when diverged from other coordinated advances in a retail store. In case any extended number of breakdowns takes orchestrate, general society could with no trouble double-cross topological clinical administrations contraptions and keep it together with their maltreatment for the presence or numerous years [6].

Detachment Violations

A next direct pressure is ensuring buyer biodata and the responsive clinical information inside these sorts of stuff. Susceptibility in an IoT clinical consideration sorts of stuff deception clear disengagement risks; thus, these pieces of devices access all customers private clinical information. Customers and masters need to guarantee that they do not convey plain (decoded) private nuances and temporarily release IoT-related contraptions.

Also, if these provisions related to clinical consideration receipt accounts, by then customer hazard missing all clinical consideration and monetary data. As per the Identity Theft Resource Center (ITRC) [7], half of all recorded information slighted in 2013 bearing an attack on clinical administration industries [7]. For instance, single could imagine how a couple of corrupt hazarders would need for an alternative to use to rule sportspersons prosperity information through gaming systems. With the potential certifiable problem message of clinical consideration sorts of stuff accusing, there is little space for breakdown when it approaches to manage these provisions in contrast to other IoT supplies. Note that the expected deadly results of a clinical device separating decrease the failure movement of these devices when stood out from various contraptions.

Masterminded Trouble

Orchestrated trouble is too a worry. Since 5G and IoT clinical consideration kinds of stuff look the vague way of thinking perils as any further topological techniques. Burgles, examiners, and lowlifes look to use perils in IT to give infringement and cause devastation. The US Department of Homeland Security (DHS) examines two courses of action of asserted organization assurance disfigurement in clinical consideration sorts of stuff that illegal could utilize, for example, persuading insulin direct to enjoy a client or ingraining a heart plant to "pass on a deadly push of intensity" [8]. Two praised security experimenters, Jay Radcliffe and Barnaby Jack, have presentation imperfections in insulin directs, there is one of the further thoroughly set up IoT clinical consideration sorts of stuff. Radcliffe (2011) uncovered that the choice to use an insulin siphon's progressive figure would permit developers to in a roundabout way talk with the stuff as of up to 140 ft away [9].

As these contraptions have slight to no safety, he could kill the siphon or reason insulin exaggerate with just $20 evaluation of devices. This ought not to occur regardless. That these attack devices have not been far-reaching could simply mean they have not yet obtained clear in haziness advance for exchange.

Customary Disruption

Regular obstructions are insensitive than the focus on murders at any rate moreover far less conceivable. Notionally, a touch of focused malware could develop over the internet and possibly get development when set up was in a clinical gadget. Such malware could influence everyone with a weak contraption. This exceptional yet conceivable situation has emerged in business IT structures and mechanical control frameworks.

4.3.3 5G with IoT Medical Devices Risk Management

The thing and firmware exceptional web-related clinical contraptions have made in much the comparative manner as different advances as a clashing blend of disparate guidelines and ways to deal with oversee wrapping up. No brand name working biological parts, planning, explicit system, or structures association backend gets by as a broadly perceived norm for any class of coordinated clinical contraptions. Any place telephones or tablets go after a practically unassuming arrangement of standard advances (Android or Apple, Wi-Fi just or Wi-Fi, and 3G, 4G, or 5G), clinical gadget creators are slanted to gather a ton of improvements and subordinate upon the size of the contraption.

Openness is mechanized by network plans through which quick electronic information movement happens. The Internet, the world's principally remarkable affiliation, includes a monstrous number of different affiliations that change in different brand names, including size, topography, and access improvement, passing on excellent multifaceted nature to the framework. Contraption and application programming separation are generally because of the nonappearance of standard programming language immediately the business. Clinical gadgets require being guaranteed sufficiently to ensure against a degenerate, yet still effectively appeared at enough to be obtained by clinical individuals. Envision a consumer with a planned pacemaker, who sensibly needs the strictest controls, by then falls detached after heart trouble while meandering out of the nation. That consumer would require for a neighborhood authority or pressing condition clinical expert to have brief enlistment to the pacemaker, yet the patient is weakened and cannot yield that help. Various producers have masterminded reactions to this issue. Some class hard passwords that are set into the structure cannot be changed. The positive perspective is that these passwords can be held in the gadget's client handbook, in a general sense found by crisis clinical pros who may foresee that they ought to think about a patient. Pitifully, programming architects can additionally fundamentally discover the secret words and abuse them.

The U.S. Computer Emergency Readiness Team (US-CERT) starting late uncovered that various people had these dangers, noticing the default secret word "let truly proximate attacker to change contraption model and cause a dismissal of association with threatening human thriving effects" [10]. Finally, there is the substance of setting up weaknesses after they are uncovered. On the off chance that, for instance, a gadget has been purposely encased, fixing the thing or firmware is not everlastingly conceivable.

In the United States, a few producers fall switch on a longstanding worry that any change, even security patches, requires FDA re-endorsing. Regardless of how this is not mindful, fixing clinical contraptions ruins exorbitant and anomalous, as producers must advance that the fixed gadget truly accumulates all clinical expected use claims.

4.3.4 Thriving, Security, and Usefulness of 5G with IoT Medical Devices Risk Management

In addition, with security risks going with other new movements, open formed effort and responsiveness are totally basic for administering and diminishing danger. This solidifies joint effort and correspondence among controllers, comparatively as between controllers, industry, and clinical and clinical thought specialists. Several propositions will help empower improvement while confining prologue to security prospects like Stress security around the beginning, as opposed to as an ominous idea, improve private–public investment, move toward uncommon differentiation in the administrative help point of view for clinical contraptions, and introduce a free voice for people all things considered.

Build Security into Devices from the Outset Rather Than as an Afterthought

Clinical gadget makers must handle a "guaranteed by-plan" way to deal with oversee inventive work. Already, security has dependably been a hint of knowing the past. Taking into account that approach, security aces have expected to manage the crazy straightforward courses originators have taken to attempt to pack security soon. Adding safety highlights to things after their fundamental rollout is a losing fight. It is absurdly unnecessary and inadequate to try to guarantee about structures effectively in the obligation regarding end client.

From time to time, adaptability and assortment in the field type safety deficiencies as gadget heads change game plans or security highlights or join advancements. An ensured by-plan approach may solidify coordinating methods of reasoning, for example, robotized logging and checking of gadget changes in the field, to see deficiencies and better direct them. Public governments, in relationship with an industry organization, may make this ensured by-plan approach more direct by giving starting financing to an open-source, standard language programming library for clinical contraptions. Since different clinical gadget creators structure their own domestic code and they are not programming authorities their changed code will without a doubt be wasteful, unequivocal to each affiliation or experience, or flooding with security openings simply grasping be found. Such little programming practices will overall make it hard to track down and fix those bugs.

Another framework for hazard the main gathering of facilitated clinical contraptions starts with interest between the makers of gadgets and programming. Makers need to work with the security business and controllers to build up a wide danger model to follow during thing improvement, plan, and transport. This model would consider them to be clinical contraption as a stage, not a self-ruling development gadget. (The PDA is another portrayal of this model.) It would make relating industry associations around unequivocal gadget lines to consider the security of types of progress related to the gadget. The objective is to pass on a clinical contraption as a liberal stage, whereupon extra movements and associations can be added.

Make 5G and IoT Relationship

To react attainably, controllers require commitment from everybody required through clear communitarian discussions, which guarantee the controller's self-administering limit without stresses of simultaneousness with the business. For a conversation with the government, one existing model is the National Health Information Sharing and Analysis Center (NH-ISAC) in the United States. The NH-ISAC itself is likely not proper for this breaking point, as it rotates around hazard reaction, yet its ability as a convener of different associates makes it significant as a model. Makers should keep improving trades between themselves. The Industrial Internet Consortium (IIC) illustrated by Intel, IBM, Cisco, AT&T, and Microsoft is a portrayal of how organization joint effort can help open production respect while in like way sustaining security.

4.4 IoT Cyber Risk Management

Expected the main presence of digital dangers and susceptibilities of the IoT networks, developing a cutting-edge digital danger the board engineering requires an understanding of both network protection innovation and winning danger the executives' systems.

4.4.1 IoT Architecture Cyber Risk

Along these lines, each layer of the IoT design has explicit safety concerns, and for the whole engineering, interrelationships with different layers and security cycles should be reflected [11]. A progression of network safety innovation works over the building focal point permits one to give an organized and incorporated perspective on online protection in the IoT. The five-layer development of big business IoT [12] by Lee is given beneath and centers around network protection issues and arrangements at the layer level.

Recognition Layer

Albeit numerous IoT gadgets are worked to be low energy and lightweight, they additionally gather enormous amounts of natural information continuously and hence execute various strategies for energy sparing. To make dependable deductions from the information created, advances, for example, AI, are regularly used [13]. Notwithstanding, inserting PC escalated security and protection assurances into lightweight IoT gadgets has been trying because of the asset obliged ability of devices [14].

The cloning of central processors for digital assaults is one of the significant security worries on the recognition layer. For instance, clones of RFID labels can be utilized to start appropriate disavowal of administration (DDoS) assaults. Actual highlights (PUFs) have been utilized for chip confirmation and distinguishing proof just as cryptographic key generation [15]. Employing alter obstruction, framework acknowledgment, confirmation, and the counteraction of cloned gadgets, PUFs chips support security [16].

Organization Layer

The organization layer assumes a basic part in the IoT structure for the general exhibition of IoT insurance, since safe transmission of information over the organization is fundamental for the working of gadgets, handling stations, and the whole IoT framework. To distinguish assaults, make remedial strides, and track packets [17], an interruption recognition framework (IDS) is utilized. Promising outcomes are likewise appeared by half and half methodologies utilizing measurement decrease and order procedures to arrange malignant exercises on the IoT networks [18]. Somewhere in the range of 2018–2023, the organization security fragment of the network safety market is required to be the biggest part of online protection, and the developing utilization of IoT applications is a key contributing element to growth [19].

Agreement Layer

At the handling level, distributed computing and haze figuring have become a standard innovation for putting away and preparing huge scope information streams produced at the same time from enormous quantities of IoT gadgets. For idleness mindful preparing of gathered information, mist figuring utilizes network devices [20]. In mist registering, IDS can be utilized to distinguish interruption on a mist node [21].

Application Layer

Venture IoT applications [22] are normally utilized in observing and control, large information and business investigation, and information sharing and coordinated effort. For instance, brilliant well-being manages profoundly modified information and requires an elevated level of security and assurance of privacy [23], since numerous IoT applications might be claimed by outsider specialist organizations, digital assaults on these applications may influence the security of other interconnected applications [24]. Different arrangements, for example, key administration, access control, heterogeneous organization verification, private data insurance, and information security, have tended to the previously mentioned security issues [25].

4.4.2 Cyber Risk Management Review

In a scholarly world and business, there is an absence of study on IoT network protection hazard the board. Thus, this writing audit is not limited to the administration of online protection danger in the IoT. Past examinations are comprehensively grouped into network protection hazard the board strategies for prevalence and extent.

Predominance Approaches

A few covering and integral danger the board frameworks exist. The National Institute of Standards and Technology (NIST) Cybersecurity System is one of the most widely recognized network protection structures. This is a significant level and simple to peruse direct made to be customized for organizations [26]. The system comprises the structure place, the phases of usage, and the profile of the structure. The framework place characterizes five of the network safety program's jobs.

ISO/IEC 27005 has been set up by the International Organization for Standardization (ISO) and the International Electrotechnical Commission as a bunch of norms (IEC). It offers counsel and methods to chairmen for the usage and the executives of data security risks [27]. The Cyber Kill Chain® (CKC) system for seven stages/chains is a danger to the board structure created by Lockheed Martin [28]. To achieve its objectives, the model investigates what the digital aggressor will do and proposes countermeasures that the defender must take to break the chain at a beginning phase just as in later stages.

A security appraisal framework created at the CERT(R) Coordination Center (Pittsburgh, PA, USA) [29] is the Operationally Critical Hazard, Asset, and Vulnerability Evaluation (OCTAVE). OCTAVE empowers an organization to distinguish and rate key IT resources, assess dangers to those resources, analyze weaknesses and impacts, and set up the security needs to diminish the danger to IT resources. The possibility of self-course with which a little, interdisciplinary group from the divisions of an association drives the appraisal cycle of the association is at the core of OCTAVE.

Greatness Approaches

This fragment talks about a couple of articles dependent on the degree of network protection hazard the board draws near. This seems to limit the range of the investigations to the assessment of the digital danger. The Bayesian Decision Network (BDN) has been applied to the organization's security hazard board framework [30]. Various significant cycles are remembered for the structure: hazard assessment, hazard decrease, and danger approval and observing, which should be performed accurately to expand the degree of organization security.

The fundamental boundaries of ISO 27005 [31] are extended by another danger appraisal framework called AVARCIBER. A scope of exercises is trailed by the usage of the structure: dispatching hazard evaluation, characterizing and investigating resources, distinguishing network protection dangers, surveying the degree of damage to weakness resource (measurement) tuples, figuring danger, and executing countermeasures. Contrasted with ISO 27005, the entire cycle is not one of a kind. More itemized exercises and pragmatic aides are given, in any case.

A contextual investigation was utilized to check the utilization of the framework. In the structure, a money-saving advantage analysis [32] was added, finding that the NIST Cybersecurity Framework does not have a valuable monetary instrument to legitimize online protection programs. The investigation shows how money-saving advantage examination may assist organizations with choosing the most fitting degree of the NIST Implementation Tier and represents the association between the degrees of the NIST Implementation Tier and the suitable degree of consumption on online protection exercises by a business. This report, anyway, does exclude authoritative data for spending choices, nor does it explain how different network protection projects can dole out monetary assets.

4.4.3 IoT Cyber Risk Management Framework

The 5G and IoT framework for digital danger of the board is recommended to help IoT security chiefs manufacture a financially savvy hazard of the executive's system for network safety. The proposed system characterizes and structures central point that influences online protection hazards into four layers, with the end goal that the hazard board exercises are organized and dissected layer by layer without dismissing network safety issues in the 10,000-foot view. An association perceives and comprehends the elements and elements of its partners through the assessment of the biological system segments. The IoT digital biological system layer tracks and surveys the climate much of the time or potentially on a progressing premise and conveys results to the next important levels. The organization investigations the present status of the network safety framework at the IoT cyberinfrastructure layer by inspecting the jobs and commitments of people, authoritative techniques, and execution of IoT online protection advancements. The IoT digital danger appraisal layer characterizes, evaluates, and organizes digital dangers and effects, distinguishes IoT resources and offices, weaknesses, and digital dangers, and assigns assets to different IoT online protection activities.

IoT Cyber Ecosystem Layer

The IoT digital environment is comprised of partners connecting cooperatively and seriously with IoT frameworks. IoT online protection innovation makers, outer clients/clients, adversaries, governments, and normalization associations are the partners of the IoT digital biological system. A move in the IoT digital biological system requires quick consideration from network protection supervisors to devise powerful security reactions for IoT gadget well-being.

To create IoT online protection techniques, numerous IoT heads do not have sufficient abilities. IoT chiefs likewise should know about the advancements in the gatherings of designers of IoT digital innovations. Designers of network safety innovations outfit IoT gadgets with the most recent apparatuses to make sure about information transmission, abstain from hacking, and maintain protection standards [33]. New innovation progressions, for example, IoT frameworks for 5G, serverless, and edge/haze processing should have the option to give better safeguard against different digital assaults on security. AI and AI will likewise exhibit colossal potential for IoT frameworks to give ongoing counteraction, distinguishing proof, and recuperation ventures with more prominent exactness than ordinary methodologies.

Foundation Layer

Associations must examine their current IoT digital foundation from both specialized and administrative perspectives while planning a future-situated digital danger the executive's procedure. Numerous examinations focus on the innovative parts of danger the executives to make sure about IT resources and administrations, yet the nonappearance of online protection uphold from associations and purchasers is as destructive to network safety as the nonattendance of network protection advances.

The conducting part of the framework is given by the staff/inside clients. This viewpoint centers around factors affecting client fulfillment, outlook, reason, and real activities on the utilization of network safety innovations and consistency with security and protection strategies. The advancement of network protection information and best practices is essential for network protection in the board's success [34].

More than 85% of senior heads concur that worker disappointment is one of their most prominent data security dangers as per an investigation led by Shred-it (2018). An experimental examination demonstrates that one reason why digital assaults happen in IoT networks at a quick movement is frail consistency with data security arrangements coming about because of social issues and the absence of information on security [35].

Senior administration upholds, security strategy, venture weakness, security preparing, and other authoritative exercises are the focal point of the hierarchical angle. These exercises are basic for improving the exhibition of IoT network safety, as these exercises are known to positively affect the perspectives and practices of IoT administrations' inside clients and laborers, just as outside clients and clients. Authoritative availability to manage digital dangers has become a significant piece of the administration of business risk [36]. This present layer's network protection innovation applies to inside properties of digital innovation.

The designers of IoT network safety innovation in the IoT digital climate are profoundly significant and can help the association in distinguishing and improving solid IoT online protection advancements. Inward network protection advances should uphold the general network protection destinations of the organization just as the targets of IoT cybersecurity. Hildebrandt recognizes three significant sorts of network safety technologies [37]: innovation that secures data secrecy; advancements that distinguish and address online dangers and weaknesses; and advances that identify and counter cybercrime.

Assessment Layer

At the point when their current IoT digital foundation is recognized by the association, they have to decide their current and potential digital danger. Three key undertakings are remembered for the IoT digital danger evaluation layer: hazard acknowledgment and hazard measurement.

IoT resources, weaknesses, and digital threats [31] are recognized by the danger ID point. Its weaknesses and danger types are characterized for each IoT resource.

IoT hazard location requires seeing how cyberattacks are dispatched by interlopers. There are two particular outlooks for gatecrashers: explorative and exploitative [34]. Interlopers generally utilize purposeful and instinctive ideas in the exploratory stages and depend on extreme experimentation. They depend on an exploitative attitude to accomplish their objectives once admittance to a framework is effective.

The effect, frequencies, and assurance probability of each digital danger classification of IoT resource weaknesses are determined at the danger measurement level. To gauge changes, the accompanying techniques and instruments are utilized. The recurrence of cyberattacks alludes to the assessed number of cyberattacks in a given timespan for hazard measurement. With chose measurements, for example, the measure of traded-off information records as well as monetary harms, the possible impact of a network protection assault over a given timespan can be determined.

4.5 Smart Home Applications Risk Management

To assist sagacious with homing advancement, a creating number of electronic contraptions and normal things are serving to improve. Considering an authentic worry for energy safeguarding and home security, new things seem to continue seeming accessible continually. Grievously, similar to all forefront mechanical degrees of progress, considering the way that our obviously splendid world goes with a couple of threats, we may all vibe creating torture or conceivably a desire to ingest data.

4.5.1 Appreciate the Smart Home Risks

With extraordinary information about the possible shrouded perils, close by your care in making philosophy for solid desire and gatekeeper, you can appreciate all the reimbursement of adroit home development with no worry.

Clever Home Technology Precautions

There are a couple of focal points of Smart Home systems, yet there are a couple of bothers. Since it is possible that producers will have the choice to ease the limitations and make splendid home development not so much perplexing but rather more secure to execute as the development creates and propels.

Opening adventure: Judge against typical home advancement, splendid home development needs a higher hypothesis.

Power floods: According to the Insurance Information Institute, you need to ensure your home is adequately guaranteed against power floods.

Compatibility: There are numerous splendid home contraption fashioners and not the total of their machines are practical. You need to use a pro to make a structure that joins all devices together, dependent upon your own understanding of development.

Internet disillusionment shortcoming: If the internet goes down, sharpest structures and devices have some sort of support that enables them to continue running. You may in any case hazard losing a segment of the further evolved functionalities of your savvy devices before closeness is restored. (Source: socialmonsters.org).

Three Potential Dangers Related to Clever Home Development Are

Smart Home Systems Leave You as Unapproved

Google's home helper, close by others, life ensures for involved specialists to get more straightforward. These devices are enticing, whether or not you are working interminably from home or dispatching your association from your kitchen. Be cautious, regardless, as they can leave your home feeble against unapproved access that could cause expensive mistakes or more horrendous. All around, all banking and vendor information is given by customers of help contraptions, allowing the structure limitless access. In case one of your housemates or youngsters chooses to orchestrate something you have not attested or all the more terrible, you may end up going up against some costly results in case someone hacks your PC.

Devices Can Leave You as Vulnerability

As terrible everything being equal to consider, the very devices you rely upon could become liabilities without a great deal of careful steps. There are numerous smart applications that grant points of interest to execute a grill on or, to jolt or open a portal, or to turn on or off external lights. A portion of the time the favorable circumstances to play out these activities are collapsed into one pack of approvals that you ought to circumspectly screen and control. To ensure that each action, close by its grouping of requests, is completely disengaged for ideal control and affirmation, you can feel that it is safer. Moreover, guarantee that you simply have the amount of uses that address the necessities of your home to thwart chaos or flood your IoT with nuances that cannot be properly organized.

Shortcoming Devices Will Leave You as

As hateful everything considered to think, without a ton of protections, the very contraptions you rely upon might become liabilities. There are a couple of splendid applications that offer the upside of murdering an oven on or shooting, or opening a passage, or turning external lights on or off. Consistently, the rights for doing these exercises are collapsed into one pack of approvals that you need to follow and direct mindfully.

4.5.2 Two or Three Key Ways to Deal with Stay Before Extortion Chances Include

Remain Safe, Stay Safe...

Insightful homes use web-related gadgets that can self-governing flick on your lights, open doors, or even alert you when you are missing the mark on milk, for instance, light switches and coolers. In any case, this comfort and receptiveness incorporate some critical destruction. Explicitly astute home structures are defenseless against a lot of security shortcomings that on the off chance that you are not mindful can place your data or property at serious risk. The most startling threats looked at by your insightful home gadgets are before you and how to restrict them.

Directed Attacks

Splendid home devices hold a hidden bonanza of individual information, from the first experience with the world date to Visa nuances that cybercriminals can take through hacking if the contraptions need good protections to demolish attacks. They would then have the option to use the taken data to dispatch zeroed in on attacks to catch you into dark game plans. For example, if a developer were to uncover that one of your family members has commitments, he could send you a phony requesting for money to help that relative. To do whatever it takes not to transform into an advanced loss, do not share financial information, for instance, bank nuances, with sharp devices, and make an effort not to relate email clients, plans, and various applications that contain sensitive information with your contraptions.

Data Fraud

From time to time, instead of breaking the splendid device of a lone individual to get his data, a developer would invade a sharp device association's informational collection to proper all of its customers' data. The data of multitudes of customers of such sense devices can be subverted by huge data breaks. A mechanized punk may apply for charge cards using your name, take out a home credit under your name, or regardless copy you and flip around your life on the off chance that you are one of those dreadful customers. Through exchanging limited individual nuances with splendid devices and ceaselessly reviewing your credit report for negative changes, you will reduce the threat of a data enter influencing you.

Secret Word Exploitation

Over and over with a feeble mystery key, quick home place focuses that interface all the sharp devices on your association are ensured about. This security pass enables splendid developers to easily invade and get to and adjust the middle point and otherwise contraptions in your home; software engineers may, for example, startle contract holders by turning on and off lights or changing TV channels. Routinely, make capricious, novel passwords and use two check levels for all your adroit contraptions to prevent virtual break-ins.

Area Tracking

To keep your concealed nuances hidden, explicitly information about where you live, you trust in your astute home contraptions. Regardless, by leaving behind your zone and making it attainable for developers to discover you and spy on you dynamically, these contraptions may sell out the assurance. For example, while tapping on a vindictive relationship from a wireless or PC, dubious smart speakers have been deceived in the past to uncover the zone of a customer straightforwardly down to the street address when the phone or PC was related with a comparative association. It will assist with calming about your bearings by evading questionable affiliations and keeping up a second association only for splendid home contraptions.

Home Intrusions

Despite the way that the potential for meddling with distant contraptions is disturbing, conversely, with the possibility of a real break-in presented by security devices, for instance, sharp gateway locks and surveillance cameras, it pales. In any of these devices, undisclosed prosperity stipulations may permit software engineers to disable cameras or open approaches to give partners access, burglarize your home, or even lock you out of your own home. Presenting security gateway underpins and security ready that is not savvy can help guard against interferences.

Machine or Property Damage

Encroachment of smart contraptions that administer fundamental home limits, for instance, cooling and warming, can be considerably more disastrous. It might be dabbled with by a software engineer with permission to your indoor controller, throwing your HVAC contraption into overdrive and making it crash. More horrible still, when you are away from home, a software engineer could turn up the oven and cause a house fire. In case you select these contraptions, look for those that license you to make sure about the settings solicitation to make it all the more difficult for anyone to transform them.

Rebel Recordings

Do you really feel like you are snooping on someone? Your tension is a lot of setup if you own a splendid speaker. Not solely can these robotized voice colleagues adequately listen to you when on; anyway, software engineers may in like manner abuse security shortcomings to break into the speaker and give their own solicitations or catch past accounts. To guarantee your assurance, sometimes delete set aside accounts, do not join security contraptions, for instance, cameras or door locks with the speaker, and consider killing your smart speakers when not being utilized.

Outsider Flaws

Numerous keen contraptions grant property holders to remotely turn on and off lights or open and close parking space passages by organizing untouchable adaptable applications or smart home stages. Nonetheless, applications without secure endorsement may allow people other than you to impersonate you and work your devices if they get their hands on your phone. Besides, a few applications bundle approvals to perform methodology on the device as opposed to requiring separate assents for every limit. This could empower a software engineer to, state, remotely jolt and open your front gateway. As a last resort, use just affirmed applications and stages to control your splendid devices.

Information Manipulation

The most unpretentious keen home hijackings leave no verification at the bad behavior scene. Since the data sent by smart contraptions like printers and sharp TVs are routinely decoded, a virtual blackguard can see—and change—data accumulated by your devices. Through this system, someone needing to break into your home could, for example, override the current video feed from your surveillance camera with changed film to make an effort not to get caught. Use advanced getting sorted out checking gadgets to help alert you to questionable correspondences or methods on the association.

Obsolete Software

It can make it significantly easier for cybercriminals to hack them for threatening purposes using outdated, peril slanted applications on clever contraptions. Buying from dependable brands that attention to affirmation and subsequently setting them to redesign normally when new writing computer programs is open is the most secure way to deal with ensuring about your contraptions. You would subsequently have the latest security revives that close by known stipulations, decline the risk of malware, and keep your sharp home running effectively and safely.

4.5.3 Savvy Cities 5G and IoT Risk Management

With the limitless advances in the development and the extension of the Internet of Things (IoT), we have seen the possibility of the "splendid city" emerge in metropolitan regions over the globe. The assignment of clever city exercises is animating far and wide, with more than 26 metropolitan networks expected to get sharp by 2025. However, while there are clear central focuses, interfacing any structure, asset, or contraption to the web presents natural risks, including data infiltrates, interference of establishment, and threats to public security. These perils are also increased when the affiliations length an entire city.

Safety Risks in Smart Cities

Affiliations generally are making ways to deal with support the game plan of sagacious city advancement, and city administrators and government specialists are either executing such exercises or, if nothing else, planning to do thusly. What may happen in case someone with a malicious objective got to a city's GPS? Envision a situation where a nation-state attacked a city's splendid framework and finished the scattering of intensity. Without real organization assurance set up, these are veritable circumstances that could achieve tremendous damage and even speak to a peril to public security. Finally, managing a sharp city requires the coordination of sensors, associations, and data examination. Regardless of the immense risks natural in any related city, there are online insurance exercises that starting now exist to support the prosperity and security of a wise city.

Risk Factors

5G and IoT sharp city things will undoubtedly turn out severely eventually. The security of a sharp city particularly depends upon two key factors—the obstacles of the advancements used and how they are completed.

Advancement Limitations

Different keen contraptions are made lightweight, which implies they simply have enough figuring ability to function as they are arranged. Edge contraptions like switches a lot have limited enrolling power, which makes scrambling them a test. It is moreover certain for splendid progressions likewise similar to any device that unexpected spikes popular for programming to get old. This is a real concern when we consider how fundamental establishment relies upon these advances to work. Systems that unexpectedly spike popular for legacy programming will reliably be slanted to attacks.

Vulnerable Implementation

The brunt of a canny city's security risks falls on how wise advancements are completed all through it. How structures and contraptions are planned can coordinate whether they are vulnerable to attacks. Public-defying on the web stages like application stores, for example, can be hurt if not ensured about. Contraptions with open ports or assembling plant arranged backhanded can be helpfully found and haggled. Given the quantity of internet-related contraptions that have straightforwardly open code stores and default capabilities, decoded and ineffectually orchestrated devices can be likewise as easy to maul. Firmware similarly becomes perhaps the main factor while considering the perils. Imagine the amount of sharp contraptions passed on all through the city; without a proper and secure strategy for pushing updates to these said devices, they are left feeble against attacks.

4.5.4 Hazards and Countermeasures

The rising of the 5G and Internet of Things (IoT) revealed a wide extent of shortcomings that can be abused by advanced criminals and diverse vindictive performers. But clever metropolitan networks are expected to manufacture productivity and efficiency, and they can present veritable threats for inhabitants and experts when network security is excused. There are a dark number of potential shortcomings and ways of thinking, and indisputably the most standard attacks showed up in Fig. 4.4.

Man-in-the-Center

An assailant invades, disappoints, or spoofs correspondences between two structures. For instance, a man-in-the-center assault on a shrewd valve in a wastewater structure could be utilized to cause a biohazard spill.

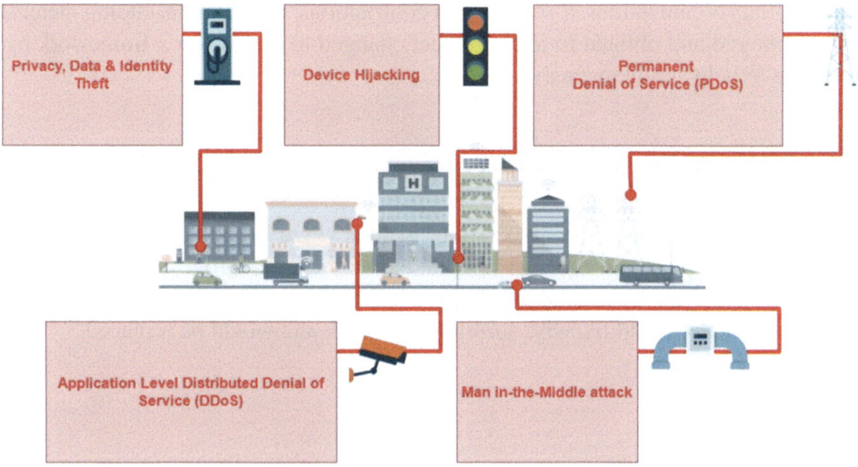

Fig. 4.4 Danger and countermeasures

Information and Rebate Deception

Data made by unprotected awesome city foundation, for example, halting structures, EV charging stations, and perception manages give progressed aggressors an adequate extent of focused individual data that can be mauled for fake exchanges and see burglary.

Gadget Getting

The assailant seizes and acceptably recognizes control of a gadget. These assaults can be difficult to perceive on the grounds that for an enormous piece of the time, the assailant does not change the fundamental accommodation of the gadget. As for a sharp city, a mechanized criminal could manhandle tied down sharp meters to dispatch ransomware assaults on Energy Management Systems (EMS) or unassumingly siphon energy from a district.

Streamed Denial of Service (DDoS)

A refusal of association assault (DoS assault) endeavors to pass on a machine or affiliation asset inaccessible to its proposed clients out of the blue or indefinitely disturbing associations of a host related to the Internet. This is regularly refined by flooding the objective with vain mentioning to shield substantial deals from being satisfied. Due to a dissipated disavowal of association assault (DDoS assault), pushing toward traffic flooding an objective beginning from different sources, making it hard to stop the electronic hostile by obstructing a solitary source. Inside talented metropolitan organizations, a great deal of contraptions, for example, ending meters, can be entered and obliged to join a bonnet changed to overpower a framework by referring to help simultaneously.

Suffering Denial of Service (PDoS)

Permanent renouncement of association assaults (PDoS), in like way suggested energetically as plashing, is an assault that harms the contraption so gravely that it requires substitution or reinstallation of stuff. In a canny city condition, a laid hold of stopping meter could likewise surrender to attack and would be replaced.

References

1. Raimundas Matuleviciu, FundSecSys, "ModFundamentals of Secure System Modelling", 2017.
2. Rodrigo Roman, Pablo Najera, Javier Lopez, Securing the Internet of Things. Computer 44 (9):51-58
3. Yuehong, Y, Y. Zeng, X Chen, and Y Fan, "The Internet of Things in Healthcare: An Overview", Journal of Industrial Information Integration 1: 3–13, 2016.
4. Niewolny, D, "How the Internet of Things is Revolutionizing Healthcare." White paper. pp. 1–8, 2013.
5. Laplante, P.A, and N Laplante, "The Internet of Things in Healthcare: Potential Applications and Challenges." IT Professional, 2–4, 2016.
6. Michael Compton and Kevin Mickelberg, "Connecting Cybersecurity with the Internet of Things," PricewaterhouseCoopers, October 15, 2014.
7. Meg Whitman, "10 Big Tech Trends in Healthcare," HP Matter, January 7, 2015.
8. Jim Finkle, "U.S. Government Probes Medical Devices for Possible Cyber Flaws," Reuters, 2014.
9. The Healthcare Internet of Things Rewards And Risks Atlantic Council Brent Scowcroft Center on International Security in partnership with Jason Healey, Neal Pollard, and Beau Woods.
10. US-CERT, "Vulnerability Summary for the Week of August 11, 2014", 2014.
11. Bendavid, Y, Bagheri, N, Safkhani, M, Rostampour, S, "IoT Device Security: Challenging A Lightweight RFID Mutual Authentication Protocol Based on Physical Unclonable Function". Sensors 2018, 18, 4444.
12. Lee, I, "The Internet of things for enterprises: An ecosystem, architecture, and IoT service business model", Internet Things Eng. Cyber Phys. Hum. Syst. , 7, 2019.
13. Hejazi, D, Liu, S, Farnoosh, A, Ostadabbas, S, Kar, S, "Development of use-specific high-performance cyber-nanomaterial optical detectors by effective choice of machine learning algorithms", Mach. Learn. Sci. Technol., 1, 2020.
14. Mollah, MB, Azad, M, Vasilakos, A, "Security and privacy challenges in mobile cloud computing: Survey and way ahead", J. Netw. Comput. Appl., 84, 38–54,2017.
15. Gao, Y, Ranasinghe, D.C, Al-Sarawi, S.F, Kavehei, O, Abbott, D, "Emerging physical unclonable functions with nanotechnology", IEEE Access, 4, 61–80,2016.
16. O'Neill, M, "Insecurity by design: Today's IoT device security problem", Engineering, 2, 48–49,2016.
17. Hodo, E, Xavier Bellekens, X, Hamilton, A, Dubouilh, PL, Iorkyase, E, Tachtatzis, C, Atkinson, R, "Threat Analysis of IoT Networks Using Artificial Neural Network Intrusion Detection System". In Proceedings of the 2016 International Symposium on Networks, Computers and Communications (ISNCC), Yasmine Hammamet, Tunisia, 11–13, pp. 1–6, 2016.

18. Pajouh, H.H, Javidan, R, Khayami, R, Dehghantanha, A, Choo, K.R, A Two-Layer Dimension Reduction and Two-Tier Classification Model for Anomaly-Based Intrusion Detection in IoT Backbone Networks. IEEE Trans. Emerg. Top. Comput. 2019, 7, 314–323.
19. MarketsandMarkets, "IoT Security Market Worth $35.2 Billion by 2023", 2019.
20. Sohal, AS, Sandhu, R, Sood, SK, Chang, V, "A cybersecurity framework to identify malicious edge device in fog computing and cloud-of-things environments", Comput. Secur., 74, 340–354,2018.
21. Yi, S, Qin, Z, Li, Q, "Security and Privacy Issues of Fog Computing: A Survey", In Wireless Algorithms, Systems, and Applications, Proceedings of the WASA, 2015.
22. Lee, I, Lee, K, "The Internet of things (IoT): Applications, investments and challenges for enterprises", Bus. Horiz., 58, 431–440,2015.
23. Almulhim, M, Zaman, N, "Proposing Secure and Lightweight Authentication Scheme for IoT Based E-Health Applications", In Proceedings of the 2018 20th International Conference on Advanced Communication Technology (ICACT), 11–14 February 2018, pp. 481–487.
24. Tweneboah-Koduah, S, Skouby, K.E, Tadayoni, R, "Cyber security threats to IoT applications and service domains", Wirel. Pers. Commun. Int. J., 95, 169–185, 2017.
25. Puthal, D. Nepal, S. Ranjan, R. Chen, J. Threats to networking cloud and edge data centers in the Internet of Things. IEEE Cloud Comput. 2016, 3, 64–71.
26. NIST, "Cybersecurity Framework", 2018.
27. ISO/IEC, ISO/IEC 27005:2018(en), "Information Technology Security Techniques Information Security Risk Management", 2018.
28. Lockheed Martin, "Cyber Kill Chain", 2009.
29. Alberts, C. Dorofee, A. Introducing OCTAVE Allegro, "Improving the Information Security Risk Assessment Process. Managing Information Security Risks: The OCTAVESM Approach" Addison Wesley, 2002.
30. Khosravi-Farmad, M, Ghaemi-Bafghi, A, "Bayesian Decision Network-Based Security Risk Management Framework", J. Netw. Syst. Manag., 2020.
31. Rea-Guaman, AM, Mejía, J, San Feliu, T, Calvo-Manzano, JA, "AVARCIBER: A framework for assessing cybersecurity risks", Clust. Comput., 2020.
32. Gordon, L.A, Loeb, M.P, Zhou, L, "Integrating cost–benefit analysis into the NIST Cybersecurity Framework via the Gordon-Loeb Model", J. Cybersecur, 6, 2020.
33. Thomas, M. 13 IOT security companies you should know. 2019. Available online: https://builtin.com/internetthings/iot-security-companies-startups
34. Esteves, J, Ramalho, E, de Haro, G, "To Improve cybersecurity, think like a hacker", MIT Sloan Manag. Rev., 58, 71–77, 2017.
35. Jeremiah, P, Samy, G.N, Shanmugam, B, Ponkoodalingam, K, Perumal, S, "Potential Measures to Enhance Information Security Compliance in the Healthcare Internet of Things In Recent Trends in Data Science and Soft Computing", Proceedings of the 3rd International Conference of Reliable Information and Communication Technology (IRICT 2018), Kuala Lumpur, Malaysia, 23–24 July, 2018.
36. Bodeau, D, Graubart, R, "Cyber prep 2.0: Motivating Organizational Cyber Strategies in Terms of Threat Preparedness" Tech. Rep. 2017.
37. Hildebrandt, M, Balance or trade-off? Online security technologies and fundamental rights. Philos. Tech., 26, 357–379,2013.

Chapter 5
Machine Learning IDS Models for 5G and IoT

Kumudavalli, Thenmozhi Rayan, and S. C. Sandeep

5.1 Introduction

5.1.1 Internet of Things (IoT)

The man-to-machine and machine-to-machine communication has shown a drift exponentially, making the operational technology gear up to connect the entities across the globe. This traditionally has defined IoT as "a pervasive and ubiquitous network that prompts the controlling and monitoring of physical environment by collecting, processing and analysing the variety of data generated by smart devices".

In general, IoT is an encompassing nonpareil situation that includes machine-to-machine and machine-to-human, RFID, location-based service, augmented reality, robotics lab-on-chip and so on, the addiction and adoption rate of the IoT is at a much faster pace as compared to any analogy devices. An added factor for the growth of IoT is a significant shift from handheld devices like palmtop/tablets based on IPV4, which is Information Technology, to an operational technology based on IPV6, that is, consumer-based internet structures to machine-to-machine communication which includes clustered systems, sensors, intelligent objects and so on. This fast-growing trend is an encouraging and highly required factor to adopt higher communication technologies in terms of the internet's capabilities to handle billions of devices [1, 2].

As IoT is an multi-user, multi-owner, shared platform and it requires equal and open access to consumers for data access with proper control on the privacy and

Kumudavalli (✉)
Dayananda Sagar College of Arts, Science and Commerce, Bangalore, India
e-mail: kumudavalli@dayanandasagar.edu

T. Rayan · S. C. Sandeep
Dayananda Sagar College of Engineering, Bangalore, India

S. Velliangiri et al. (eds.), *Secure Communication for 5G and IoT Networks*, EAI/Springer Innovations in Communication and Computing, https://doi.org/10.1007/978-3-030-79766-9_5

exclusivity of data being shared. This is the primary concern to properly adopt complex security norms leading to a thorough study on security aspects of IoT [3].

IoT has multiple challenges with security aspects. It has to authenticate multiple networks, make data available to different collectors, care about the privacy between multiple consumers, continue service for data availability and handle the unknown risks. The critical processes must be ensured of security, that is, the attacks that may cause damage to the data or process. Machine Learning algorithms are playing an important role in IoT security [4]. The major implemented solution for several attacks is the intrusion detection system (IDS). Further in this chapter, we learn specific ML algorithms that significantly impact IoT security [5, 6].

5.1.2 5G Technology

Worldwide accepted and implemented broadband cellular network is the fifth generation technology in the field of telecommunication which found its peak in the year 2019. The study predicts the successor of the 4G, that is, the 5G will have approximately 1.7 billion users by the next 5 years. 5G technology uses the concept of cell; which are the small geographical areas often known as the service areas. Radio waves are the medium through which all the wireless gadgets within the cells are connected to the telephone and internet network [7, 8].

5G technology is a boon in terms of bandwidth, which enhances the download speeds may be up to 10 Gbps. Due to its higher bandwidth, the 5G network will not serve the existing cell phones but also can be used as general internet service providers for other computing devices like desktops and laptops. This technology is at the forefront against the existing ISPs like cable internet, which is also ruling the domain of machine-to-machine areas and IoT [9]. The efficiency of the 5G networks is achieved by using higher frequency radio waves through smaller geographical cells. Figure 5.1 gives a clear understanding of the various frequency bands the 5G network is being divided into with its performance indicators.

The 5G network also has limitations in terms of quantity of cells required, implementation process and cost incurred, the transmission of signals through specific materials, and so on. In spite of all these, 5G is set to achieve its speed which could be handled during the rollout process. 5G is the tailor-made technology for artificial intelligence and machine learning domains as it supports low latency and very high speeds enabling a heterogeneous mix of traffic from a huge number of devices. AI/ML aggregates by learning from complex patterns to provide scope for self-governing operations converting the 5G to be a scalable real-time data-driven network.

Optimization, security, fault prediction and fraud detection are an integral part of any network service. AI and ML are being used in 5G networks for planning, automating the operations of networks, cost optimization and improving service and customer experience. 5G RAN, Integrated Access Backhaul (IAB) and distributed

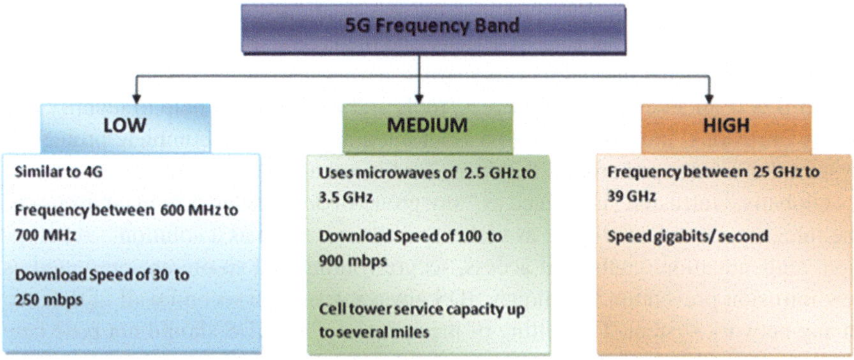

Fig. 5.1 5G frequency bands

cloud layer (5G Edge/Core) are all having the features of AI/ML being implemented [10].

ML algorithms and modules are encouraging the researchers to share their views, opportunities and challenges to be addressed in the operational domain of 5G. AI emancipated by machine learning and virtualization is qualifying efficient, secure, reliable and cost-effective network.

5.1.3 Intrusion Detection System (IDS)

With growing challenges and changes in the techniques in the tools being used by the attackers, the Intrusion Detection System (IDS) is finding its pinnacle position as an implemented solution against injurious attacks.

The rapid growth in the field of Information Technology in terms of smart devices and connectivity concerns has put forth a variety of challenges to the infrastructure domain. Building and maintaining a reliable network is an everyday challenge as the field of IoT is having a billion device turnover, that is, the exponential increase in the interconnected smart devices are prone to challenges including authentication, authorization, security, connectivity and speed. Since the active devices in an IoT environment are big in number, the output data is also huge, which needs to be handled with care. The big data or the huge amount of data through IoT is more risk-prone. The attacks on these output data are of different types and are active at different stages. Recognizing, handling and preventing these attacks are evergreen tasks of the network administrators. The IDS is one of the best proven methods to detect and prevent the networks from vulnerabilities [11].

The intrusion is an unapproved activity within the network that takes place actively or passively. In a secured environment, the first step is intrusion prevention and if that fails, then intrusion detection comes into role. It is the process of detecting any suspicious activity or behaviour in a network by the internal or network members. Generally, the IDS provides the information regarding the intruder, its location, time of intrusion, type of activity and the layers, that is, physical/data link/network in which the intrusion occurs.

An IDS is an assembly of resources, tools and methods that help to identify and notify intrusions. Intrusion detection is an integral part of the complete protection system that is wrapped around the device and is not a single thread measure.

Globally, "intrusion" is defined as "any group of actions that acts to compromise the integrity, confidentiality or availability of the resource as a solution"—encryption, authentication, controlled access, secure routing and so on, are presented as few intrusion prevention techniques. IDS always act as the second wall of defence in any network system. The setting of the process of any IDS should not pose new challenges to the security systems, namely, introducing of newer weak points, demand more system resources, new overheads, not being transparent and active, showing non-cooperative moves by not using standards, not reliable and so on.

The IDSs can be majorly classified into different classes/groups based on the central component data. The IDS are used based on data storage, its analysis methods and their behaviours like being active or passive.

Figure 5.2 gives a brief classification of IDS based on data storage, analysis and response modes.

The detection methodologies can be signature based or anomaly based. The anomaly based detection follows the statistical behaviour modelling. In this case, the normal functioning of the members are profiled and some amount of deviation from the normal functioning or behaviour is called an anomaly. This method is set to detect intrusions in a very accurate and consistent way, keeping the rule that the network is studied under the static behaviour patterns. The advantage of anomaly based detection is that it is best suited to detect unknown or no previous history attacks.

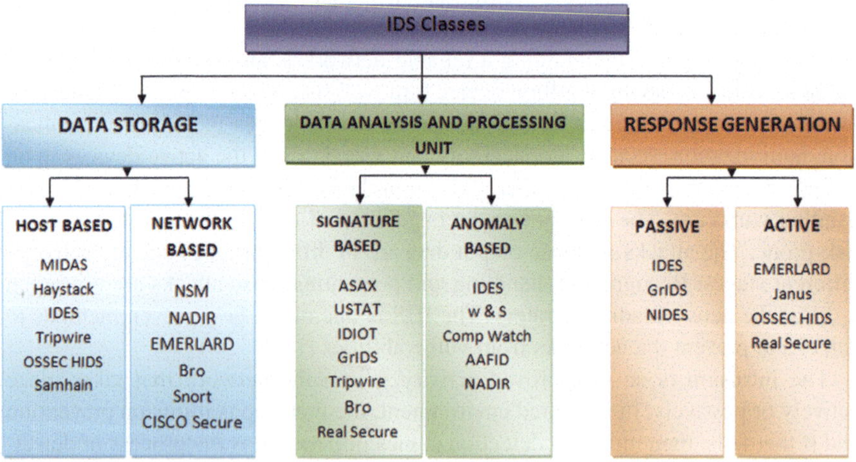

Fig. 5.2 Intrusion detection system classes

The statistics based time interval study of the network usage is the baseline of the anomaly based intrusion detection system, which is effective in the detection of unknown attacks. This forms the base for introducing the machine learning methods to build models simulating the regular activities based on the existing database and then to train and compare the new behavioural patterns with the existing active model.

5.2 Machine Learning and IDS

The intrusion detection systems traditionally work to prevent and detect intruders [11]. The IDSs also have a fallback in terms of few limitations, namely, many real attacks are overlooked due to the large number of false alarms raised within the system. The version control or regular updates pertaining to the software development enforces the IDSs to keep up with the new threads. They are not equipped to handle protocol-based attacks. Detection capacity outside the network is a challenge. The network IDS can be a hurdle as the inbound and the outbound traffic has to pass through it. As the IDS databases majorly rely on audit logs, any modification of these logs results in the threat of integrity (Host IDS).

Considering these limitations, the IDSs capabilities can be enhanced through machine learning models, which follow a thumb rule of learning and improving from experience.

Machine Learning is an integral part of the artificial intelligence domain, which caters to the needs of advancements in the technology domain. It is an exclusive field of computers that increases the capabilities of any system to learn and improve from experience without being explicitly programmed. The major advantage of the machine learning model is to develop programs and algorithms that can discover themselves. The process of learning starts with existing data or observations, which have patterns and to look for these patterns and make better predictions based on the set of examples. This feature of ML finds huge demand in the field of IoT as the numbers of interconnected devices are increasing rapidly allowing the data to grow and creating a huge number of examples to learn from.

Generally based on the data set the machine learning algorithms are categorised as "Supervised", "Unsupervised" and "Semi-supervised". The supervised ML algorithms are capable of predicting future events based on the existing data or past events. It uses classified/labelled data for the predictions. The data set used is generally called a training data set, as they are used to make the machine understand and be equipped to make predictions about the output values.

On the contrary, the unsupervised machine learning algorithms play a role in the context of training the unmarked or non-classified data. It is basically the set of algorithms used to infer a function to describe the hidden structures from the unlabelled data. They follow patterns, similarities and differences to group the unsorted data without any pre-trained data.

Semi-Supervised learning is a mediate between supervised and unsupervised learning. They are used when the collection of data set for training neither has enough labelled data nor has all to be unlabelled. The semi-supervised techniques can be used to increase the volume of the training data, as it is a blend of few labelled data and a lot of unlabelled data.

The unsupervised algorithms can learn faster from patterns of the networks and can give a signal of anomaly without having a proper labelled data set. They can detect new intrusions attacking the network easily but are prone to false alarms as they are the most occurring patterns. Hence only an unsupervised K-Mean algorithm is majorly used for the study of unlabelled data. To have an optimal study or to predict the future events efficiently, which can reduce the false positive factors, it is advised to use the labelled data and to build a supervised machine learning model. This can be achieved by training the model well to differentiate between normal and a dangerous attack packet in the network. This enables the model to handle the attacks and is also capable enough to recognise the discrepancy of those attacks [12, 13].

Many supervised and unsupervised algorithms are in practice, namely, K-Mean Clustering, Artificial Neural Network, Bayesian Network, Random Forest, Random Tree, Decision Table, Markov Model, Support Vector Systems and many more.

Further this chapter deals with a thorough literature study of various Machine Learning algorithms and models which are showing apt results in the field of IoT and 5G networks. Since the fifth-generation telecommunication revolution has given realistic challenges to the networking world and to which the added wing IoT is also creating a promising growth curve in terms of connectivity, production and scalability. It is a must for the information technology field to enhance and equip to be able to handle the exponential growth and change.

5.3 Literature Review of Algorithmic Approach to IoT and 5G

Jordan Lam and Robert Abbas (2020) have proposed an "SDS (Software Defined Security)", which is a means of creating a mechanized, lithe and scalable network defence structure. According to their study, the SDS is capable of supporting recent inclusions in machine learning to build a "CNN (Convolutional Neural Network)", which makes use of "NAS (Neural Architecture Search)" to detect the inconsistent network traffic. They have also mentioned the SDS application to 5G network as it is efficient enough to handle the intrusions, that is, detection of the intrusions in the system that can define a more proactive and end-to-end defence. To verify this supposition authors have considered regular and abnormal network flows from a simulated situation and have analysed the same with a CNN. The results obtained from this process are high as the model was successful in identifying "benign traffic with a 100% accuracy rate and anomalous traffic with a 96.4% detection rate". This shows the efficiency of the system and also gives a workable opportunity for the discovery of encrypted dangerous/infected network traffic.

Shahadate Rezvy et al. (2019) have implemented "deep encoded dense neural network" algorithm to 5G and IoT domain for detecting and analysing various forms of attacks. They have applied and analysed the method using the "Aegean Wi-Fi Intrusion" data set. The outcomes have shown an exceptional result giving an overall recognition accuracy of about 99.9% for "Flooding", "Impersonation" and "Injection" kind of attacks. The authors also have given a detailed description of various approaches mentioned recently in the literature with their experimental algorithm which has shown a significant enhancement with regard to accuracy and detection pace. They have evaluated the model using a confusion matrix and they have achieved a "true positive rate (TPR)" of more than 99% for every attack category and normal association. For the attacks due to flooding, only 1.2% of false predictions were made by the model, that is, only 177 wrong predictions were made out of a host of 14k instances. They found an even lower rate of false prediction with 0.1% and 0.9% for the other attacks, that is, injection and impersonation respectively [14, 15].

Irin Anna Solomon et al. (2019), in their article Intrusion Detection System Using Deep Learning, have discussed the security aspects in the system administration that can use the IDS to eliminate unauthorized access. They have adopted the Deep Learning technique for efficient detection of intrusions and their activities that can cause damage to the network. They have used the Feed-forward approach on KDD99-CUP data set to compare their methodology with the existing related works. In the first phase of data pre-processing, they have removed the redundant values by reducing the data points to 145,586 samples. Later they found results are graphically explained. The authors have used tensorflow as their programming approach to obtain the results and it has shown an outstanding accuracy for the proposed system. The results have demonstrated that the proposed approach offers high levels of accuracy and precision along with reduced training time [16–18].

Ali Azawii Abdul Lateef et al. (2019) have given an extensive survey on various Machine learning, particularly, the deep learning approaches available for the study of networks using IDS. They have classified the algorithms under study into three categories. That is, discriminative, hybrid and generative. Based on these types they have also given a detailed description of different data sets and frameworks related to the study, namely, DARPA, KDD99 and NSL-KDD. They have emphasized the growth of newer algorithmic approaches to handle the fast changing tools and techniques that are being used by the attackers. They have concluded that AE and RNN are performing better compared to CNN as CNN uses image data set to show a better performance also suggested on using a hybrid approach for feature selection and feature extraction as it increases the accuracy of intrusion detection [19, 20].

Razan Abdulahaammed et al. (2019) have used two feature dimensionality reduction approaches, namely, "Auto-Encoder (AE) and Principal Component Analysis (PCA)". Then the consequential low-dimensionality feature obtained from the two methods is then used to construct different classifiers like "Random Forest", "Bayesian Network", "Linear Discriminant Analysis" and "Quadratic Discriminant Analysis" for scheming IDS. Their tentative results considering all low-dimensional features with respect to binary and multi-class classification demonstrate improved presentation in stipulations of detection rate, false alarm and accuracy. They have used the CICIDS2017 data set for experimentation and have obtained an accuracy rate of 99.6% [21, 22].

JooHwa Lee and KeeHyun Park (2019) have proposed "high-performance network intrusion detection system" based on deep learning aptly designed for a situation where considerable, noteworthy imbalance exists between benign and irregular traffic. They have used "autoencoder (AE)" and the "generative adversarial networks (GAN)" model throughout deep learning which, are unsupervised learning methods to resolve the disparity of data and intrusion findings with high performance. The "AE-CGAN (autoencoder-conditional GAN)" model has been implemented to progress the performance capacity of intrusion detection. The authors claim that their model has high efficiency in comparison to the uncommon classes based on the "GAN model" with an intention to improve the performance deprivation caused by data imbalance after considering the distinctiveness of the data to a lesser level using the "autoencoder" model. For determining the performance of the "AE-CGAN" sculpts, they have differentiated the data using "random forest (RF)", typical ML classification algorithms. The "Canadian institute for cybersecurity intrusion detection system (CICIDS) 2017", data set is being used by the authors for experimentation. They have analysed the performance variations among all the three types which include "single-RF" a classification algorithm-based model, "AE-RF" based on data features and the "AE-CGAN" model used after data feature processing and data imbalance. Their study results show that the performance of the "AE-CGAN" model suggested by them was the best. Especially in the situation of unbalanced data, the authors claim that the "AE-CGAN" model showed better and accurate results of recall and F1 score, with an indicated rate of 93.29% and 95.38% respectively [23].

Clifford Green et al. (2018) have presented a qualified evaluation of "deep learning approaches" to "network intrusion detection". They have worked on three different models, a "vanilla deep neural net (DNN)", "self-taught learning (STL)" and "Recurrent Neural Network (RNN)" based "Long Short Term Memory (LSTM)" on their precision and accuracy rate. Their capabilities have been applied and making use of network intrusion data set provided by "Knowledge Discovery in Databases (KDD)". Later the results are compared with the "baseline shallow" algorithm that uses "multinomial logistic regression" to check if deep learning models perform better on the KDD data set. The authors have mentioned that their observations reveal that the "autoencoder" can recognize the attack types with an accuracy of 98.9%. But on the contrary, the "Long Short Term Memory (LSTM)" model gave a 79.2%. They have also considered the step to tune the hyper parameters, which were essential to improve the accuracy of the "LSTM" model. Since the training set depends on the predictions of these models, the class imbalance could have been the reason for lower accuracy. Furthermore, they described that the "self-taught learning model", as a result of dimensionality reduction decreased the existing features to 10 in the "autoencoder". The performance result had a higher accuracy as compared to the SMR results which was performed on the cleaned "NSL-KDD" data set, which gave a much lesser ratio of 75.23% considering the entire 41 features of the original data set. With this, the authors wrap up with the conclusion that the "autoencoder deep learning algorithm" is a good model for NIDS. They say that the "STL model" could be used to study the behaviour of unclean data [24].

Mohammad Almseidin et al. (2018) have created a breakthrough analysis report based on various Machine Learning algorithms for IDS. They have conducted several experiments to evaluate the performance factors using various machine-learning classifiers algorithms on the "KDD intrusion" data set. They have used "J48", "Random Forest", "Random Tree", "Decision Table", "MLP", "Naive Bayes" and "Bayes Network" machine learning classifiers for their study [25]. It was succeeded with many computational checks or performance metrics to bring out promising results and to examine the chosen classifiers. Their major focus was on "false-negative" and "false-positive" performance metrics to enhance the detection rate of the intrusion detection system. They performed the experimental study with Ubuntu 13.10 platform using WEKA a machine learning tool. They are of the opinion that the implemented experiments have confirmed that the "decision table classifier" has achieved the lowest value of false negative where the "random forest classifier" has achieved the utmost average accuracy rate. The experiments conducted by the authors have shown that there is no sole machine learning algorithm, which is capable enough to handle all types of possible attacks efficiently. They are at the conclusion that the "decision table" gave the least possible false negative value of (0.002), although it was away from the greatest range of accuracy detection. On the other hand, "Bayes network classifier" had the highest value for perfectly identifying the normal packets. "Random forest classifier" registered the highest accuracy rate of 93.77%.

Bandar Alotaibi and Khaled Elleithy (2016) were successful in implementing a "majority voting technique" for "Wireless Intrusion Detection System" one of its kind approach where they have worked on the misuse of Wireless LAN Intrusion Detection System (WIDS). They have tested many Machine Learning algorithms in this regard they have found promising results, which improve the accuracy and computation time. They claim that the "Extra Trees", "Random Forests" and "Bagging" were the best-performing algorithms.

Later the authors have used a "majority voting technique" to vote on these algorithms. The "Bagging classifier" and the author's proposed customized "voting technique" have good results about 96.25% and 96.32% respectively when tested on all the features. They have also extended their study by using a "data-mining technique" based on the "Extra Trees ensemble" method to find the most important features on the "Aegean Wi-Fi Intrusion" Data set. They have chosen the most 20 significant features to come to a conclusion that the "Extra Trees" and "voting technique" were the best performing classifiers in terms of accuracy with an accuracy of 96.31% and 96.32% respectively [26].

Table 5.1 gives brief details about the various Machine Learning algorithms being used for the study of IDSs in the field of Networks, IoT and 5G.

Table 5.1 ML algorithms data set and performance description

Authors	ML algorithms/ techniques used for experimentation and proposed techniques	Data sets	Performance
Jordan Lam and Robert Abbas (2020)	Proposed: SDS—Software Defined Security Used: CNN	Simulated	100% accuracy in finding benign traffic 96% in finding anomalous traffic
Shahadate Rezvy et al. (2019)	Deep auto-encoded Neural Networks	Aegean Wi-Fi intrusion data set	99% for "Flooding", "Impersonation" and "Injection" types of attacks
Irin Anna Solomon et al. (2019)	Deep Learning Technique, Feed Forward approach	KDD99-CUP	High-level accuracy and precision with reduced training time
Ali Azawii Abdul Lateef et al. (2019)	Discriminative, Hybrid and Generative Deep Learning Algorithms	DARPA KDD99 NSL-KDD	AE and RNN perform better than CNN
Razan Abdulahaammed et al. (2019)	Auto-Encoder (AE) and Principal Component Analysis (PCA)	CICIDS2017	Very good in detection rate, false alarm and accuracy of 99.6%
JooHwa Lee and KeeHyun Park (2019)	Used: AutoEncoder (AE) and Generative Adversarial Networks (GAN) Proposed: AE-CGAN (AutoEncoder-conditional GAN)	CICIDS2017	AE-CGAN's performance is better with recall and F1 scores having an accuracy rate of 93.29% and 95.38% respectively
Clifford Green et al. (2018)	"Vanilla deep neural net (DNN)" "Self-taught learning (STL)" "Recurrent Neural Network (RNN)" "Long Short Term Memory (LSTM)"	KDD NSL-KDD	98.9% accuracy using the AE method. 79.2% Long short term memory (LSTM) model
Mohammad Almseidin et al. (2018)	J48, Random Forest, Random Tree, Decision Table, MLP, Naive Bayes, and Bayes Network	KDD	"Decision table" achieved the lowest false negative value of (0.002). "Bayes network classifier" had the highest value for correctly detecting the normal packets. "Random forest classifier" gave the highest accuracy rate of 93.77%
Bandar Alotaibi and Khaled Elleithy (2016)	Used: Extra Trees, Random Forests and Bagging Proposed: Majority Voting Technique	Aegean Wi-Fi Intrusion Data set (AWID)	Bagging classifier gave 96.25% Extra Tree gave 96.31% and the best accuracy with Majority Voting Techniques with 96.32% accuracy

5.4 Conclusion

The ever-growing information technology field is a goldmine for the application field where the researchers, programmers, academicians and industry domain experimenters have a huge scope for invention, research and development. The added feather to this growth is the field of IoT which is challenging the majority of the domains with its rapid growth speed.

5G technology is the new and forefront service provider in terms of telecommunication facilities is also being considered as one of the elite factors in the technology growth curve. As the technology and the devices related to it grow the data and the security issues pertaining to data also increase. The most widely accepted and used method for detecting and preventing attacks on the network is Intrusion Detection System (IDS). The interconnected or network of devices that are constantly involved in data exchange are being monitored for vulnerabilities and various methods are also used for the same. Various Machine Learning algorithms have been used, implemented and tested on a variety of databases related to network traffic. Many researchers have made an extensive study related to the various attacks that the networks are prone to, the behaviour of the intruders, the algorithmic evaluation the accuracy factors and so on., which are essential but still are not sufficing the existing threats as the technology advancements are challenged by the variety of techniques used in the attackers' mode also. A step ahead in the algorithmic revolution was the use of Deep Learning methods to study network attacks. It is being widely used in this domain but it has not reached the real-time scenarios which are beyond the experimental conditions. So there is a huge scope for further advancements required to handle the huge traffic created through high performance computing/smart devices.

References

1. IoT: number of connected devices worldwide 2012-25, Statista: https://www.statista.com/statistics/471264/iot-number-of-connected-devices-worldwide/, 2016
2. https://tools.cisco.com/security/center/resources/secure_iot_proposed_framework
3. Okwori Anthony Okpe et al., "Intrusion Detection in Internet of Things (IoT)", International Journal of Advanced Research in Computer Science, Vol 9, No. 1, 2018
4. A. Alrawais, A. Alhothaily, C. Hu, X. Cheng, Fog computing for the internet of things: Security and privacy issues, IEEE Internet Comput. 21 (2) (2017) 34–42.
5. Securing the Internet of Things: A Proposed Framework, 2016, http://www.cisco.com/c/en/us/about/security-center/secure-iot-proposed-framework.html.
6. S. Velliangiri and P. Karthikeyan, "Blockchain Technology: Challenges and Security issues in Consensus algorithm," 2020 International Conference on Computer Communication and Informatics (ICCCI), Coimbatore, India, 2020, pp. 1–8, https://doi.org/10.1109/ICCCI48352.2020.9104132.
7. Jordan Lam and Robert Abbas, "Machine Learning based Anomaly Detection for 5G Networks", arXiv:2003.03474v1 [cs.CR], Mar 2020.
8. Ericsson, A Guide to 5G Network Security, Ericsson.com, pp. 120.

9. 5G Americas, The Evolution of Security in 5G, 5G Americas Whitepaper, pp. 140, Oct. 2018.
10. CISCO, Applying Security to the 5G world, CISCO, 16-Jan-2019. [Online]. Available: https://www.cisco.com/c/dam/m/en us/network-intelligence/service-provider/digital-transformation/knowledge-network-webinars/pdfs/0116SECURITYCKNPDF.pdf. [Accessed: 07-Apr-2019].
11. Mohammad Almseidin et al., "Evaluation of Machine Learning Algorithms for Intrusion Detection System", https://www.researchgate.net/publication/322328775, 2018.
12. Lee, C.H.; Su, Y.Y.; Lin, Y.C.; Lee, S.J. Machine learning based network intrusion detection. In Proceedings of the 2017 2nd IEEE International Conference on Computational Intelligence and Applications (ICCIA), Beijing, China, 8–11 September 2017; pp. 79–83.
13. P. Amudha, S. Karthik, and S. Sivakumari, "Classification techniques for intrusion detection-an overview," *International Journal of Computer Applications*, vol. 76, no. 16, 2013.
14. Shahadate Rezvy et al., "An efficient deep learning model for intrusion classification and prediction in 5G and IoT networks", https://doi.org/10.1109/CISS.2019.8693059, IEEE, March 2019.
15. Velliangiri, S., & Premalata, J. (2020). A Novel Forgery Detection in Image Frames of the Videos Using Enhanced Convolutional Neural Network in Face Images. Computer Modeling in Engineering & Sciences, 125(2), 625–645
16. Irin Anna Solomon1 et al., "Intrusion Detection System Using Deep Learning", Asian Journal of Computer Science and Technology ISSN: 2249-0701 Vol. 8 No. 2, 2019, pp. 105–110
17. Tavallaee, M., Bagheri, E., Lu, W., et al.: A detailed analysis of the KDD CUP 99 data set. In: 2009 IEEE Symposium on Computational Intelligence for Security and Defense Applications, pp. 1–6, July 2009
18. Tensorflow lite: A new mobile-specific library (2017). https://www.tensorflow.org/mobile/tflite/
19. Velliangiri, S., Cristin, R., & Karthikeyan, P. (2018). Genetic gray wolf improvement for distributed denial of service attacks in the cloud. Journal of Computational and Theoretical Nanoscience, 15(6–7), 2330–2335.
20. Ali Azawii Abdul Lateef, Sufyan T. Faraj Al-Janabi, Belal Al-Khateeb, "Survey on Intrusion Detection Systems based on Deep Learning", Periodicals of Engineering and Natural Sciences, ISSN 2303-4521 Vol. 7, No. 3, September 2019, pp. 1074–1095
21. Van Der Maaten, L.; Postma, E.; Van den Herik, J. Dimensionality reduction: A comparative review. J. Mach. Learn. Res. **2009**, 10, 66–71
22. Razan Abdulhammed et al., Features Dimensionality Reduction Approaches for Machine Learning Based Network Intrusion Detection, Electronics 2019, 8, 322
23. JooHwa Lee and KeeHyun Park, "AE-CGAN Model based High Performance Network Intrusion Detection System", Applied Sciences, 9, 4221; https://doi.org/10.3390/app9204221, 2019.
24. Clifford Green et al., "Comparative Study of Deep Learning Models for Network Intrusion Detection", SMU Data Science Review, Vol. 1, No. 1, Art. 8, 2018
25. Velliangiri, S., Karthikeyan, P. Hybrid optimization scheme for intrusion detection using considerable feature selection. Neural Comput & Applic 32, 7925–7939 (2020).
26. Bandar Alotaibi and Khaled Elleithy, "A Majority Voting Technique for Wireless Intrusion Detection Systems", https://doi.org/10.1109/LISAT.2016.7494133, IEEE, 2016.

Chapter 6
Industrial Automation of IoT in 5G Era

S. P. Anandaraj, S. Poornima, R. Vignesh, and Vinayakumar Ravi

6.1 Introduction

Nowadays, the improvements in business, industry professionals, and people's lives lie in the development of the Internet of Things (IoT). In these interconnected works, massive million or multi-trillions of economics are growing second by second, which focuses on market growth globally or just the point of internet access everywhere. Because of this mobile applications play a vital role in interconnecting people globally. Mobile applications are already in the market as 1G, 2G networking people by offering voice services. Moreover, 3G and 4G networks provide broadband services by connecting people over the internet quickly [1].

In this global world, many things are available other than humans and are interconnected widely. Garner, a famous Analysis Company, released a report in 2017 stating that over 8.4 billion IoT gadgets exist in society. In contrast, in 2016, just 40% of the gadgets were connected, resulting in more than 20.4 billion gadgets interconnected by 2020. IDC is another analysis company, which stated that counties had spent $772.5 billion in 2018 on gadgets manufacturing. In contrast, it is more than a 15% hike in the amount spent in 2016. The same company released a prediction report stating that nearly $1 trillion will be crossed in 2020. It will increase year by year.

Nowadays, IoT gadgets are interconnected globally via wi-fi technology, partitioned into four connection mechanisms, such as wired technology, mid-range wi-fi technology, long-range wi-fi technology, and so on.

S. P. Anandaraj (✉) · S. Poornima · R. Vignesh
CSE Department, Presidency University, Bangalore, India

V. Ravi
Center for Artificial Intelligence, Prince Mohammad Bin Fahd University, Khobar, Saudi Arabia
e-mail: vravi@pmu.edu.sa

S. Velliangiri et al. (eds.), *Secure Communication for 5G and IoT Networks*, EAI/Springer Innovations in Communication and Computing,
https://doi.org/10.1007/978-3-030-79766-9_6

Satellite technology. Each partition consists of a wide variety of distinctive techniques and standards falls. Alternatively, IoT technology can be partitioned into short-range and wide-range interconnections based on their target usage. The IoT devices are connected via many technologies such as Zigbee, Z-Wave, wi-fi, Bluetooth, and so on. As a development, the IoT gadgets are interconnected via cell technologies such as GSM, LTE, 5G, and so on. The latter used are the services provided by an unlicensed firm such as Sigfox and Lora. The cell-based technologies offered a wide variety of benefits and services to users. Due to this, the wi-fi cell-based technology remained a part of World Third Generation Partnership Project (3GPP) technology. IoT faced many challenges during the introduction of 4G LTE and overcame those challenges in 5G [2].

In the Worldwide Web, the "Massive IoT" term became very popular since it connected trillions of internet gadgets, nodes, and computers. Especially, mobile gadgets went viral due to their extended connectivity and by their various services provided to target users. This massive IoT captures a larger scale of users than 3GPP by offering one million units per kilometer [3].

In addition, the cellular networks need to guide the most straightforward gadgets efficaciously, sometimes speaking, which are ultra-energy efficient for the long term with 10-plus years' battery life. However, the big IoT does not have to wait for 5G scale deployments for success. 4G science is helping the big IoT today. According to the Ericsson Mobility Report–Q4 2018, there will be approximately 400 million IoT gadgets with cellular connections at the quit of 2016. That variety is projected to attain 1.5 billion in 2022 or around 70% of the wide-area category [4]. In 2018, cell phone connections were surpassed in numbers with the aid of IoT devices [5]. This boom is due to increased industry focal point and 3GPP standardization of mobile IoT technologies. Cellular IoT connections advantage from enhancements in provisioning, system administration, and service enablement. Perhaps even more excellent significantly, cellular networks also provide ubiquitous coverage. Unparalleled reliability, security, and overall performance are required to employ the most annoying IoT applications [5, 6]. 3GPP applied sciences such as 4G Long Term Evolution (LTE) can supply wide-area IoT connectivity—LTE is the most extensively deployed and fastest-growing wi-fi technological know-how with 3.7 billion connections global at the cease of 2018.6 With 637 LTE business networks7 worldwide as of April 2019, LTE will continue to amplify its insurance and connections over the next decade forming the basis for the subsequent technology of wi-fi technological know-how—5G. The Enhanced Machine-Type Communication (eMTC), otherwise named as LTE-M, Machine-Type Communication, and Narrowband-Internet of Things (NB-IoT), is LTE-type IoT, which 3GPP founded as an additional narrowband. Both are combined to form an LTE IoT. Both technologies serve the target users with decreased complexity, reduced power, extreme coverage, and optimized node utilization, which is better than LTE services from an existing broadband connection. These LTE Services motivated the Massive IoT to get integrated with LTE IoT. Figure 6.1 shows the overview of the increasing rate of IoT market statistics.

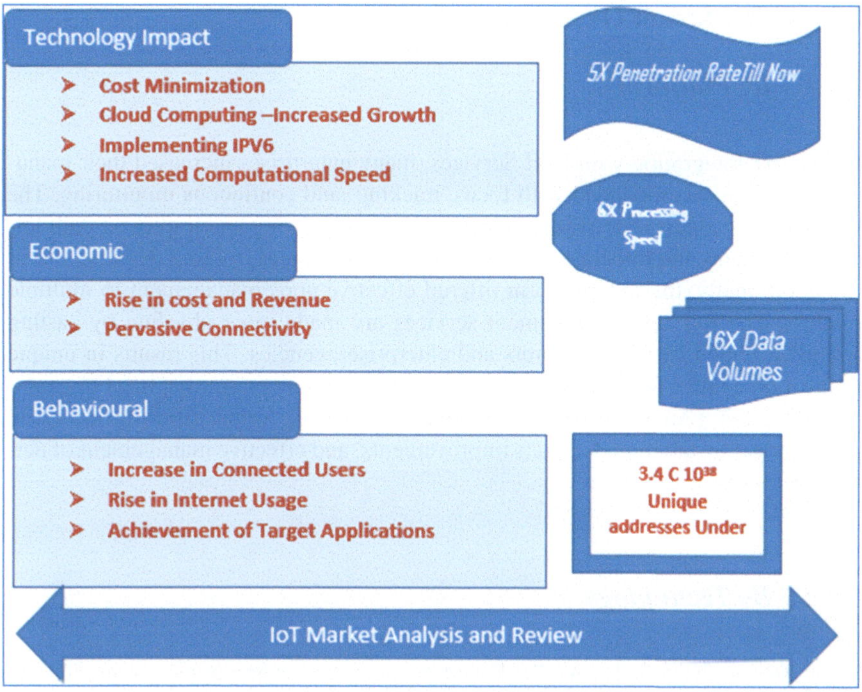

Fig. 6.1 Increasing rate of IoT market statistics

6.2 Increased Growth in the IoT Market

The LTE-MTC is abbreviated as Long-Term Evolution-Machine-Type communications started by the LTE-M enterprise as a part of LPWA Technology standards. Therefore, it started helping the users with reduced design complexity, more comprehensive connectivity, reduced power, low latency, and broader network coverage with the reusability of LTE without any implementation medication. At fewer times, LTE carrier will be standalone technology in serving multiple users via LTE-M installation [7].

The 3GPP enterprise fulfills the LPWA requirements by integrating Narrowband IoT (NB-IoT) by offering various IoT services [8]. NB-IoT's invention resulted in increased LAN connections, Increased gadgets connectivity, improved fault tolerance, reduced gadgets price, reduced power, and a well-connected network framework. The NB-IoT can be integrated with LTE carrier as "in-band," and it will act solely to serve target users. Additionally, NB-IoT can also be deployed in an LTE carrier's guard band [9].

6.3 IOT Market Overview

6.3.1 By Platform

Due to IoT integration with LTE Services, many enterprises increased their manufacturing gadgets embedded with focus, tracking, and continuous monitoring. The IoT-enabled gadgets started supporting continuous tracking and improved fault tolerances in many intermediate nodes, such as Wi-Fi devices, routing devices, and so on. Additionally, the IoT platform offered effective apps management to multiple vendors. The IoT tool management services are made more absolute by casting options invented by manufacturers and enterprise agencies. This results in unique application control formats defined by enterprise vendors presently [10]. For example, AWS IoT (Amazon IoT) is the advanced control format offering increased speed gadget installation, Gadgets improvements, and effective management of network devices.

6.3.2 By Technology

From 2019 to 2025, IoT with RFID and connectivity technologies got popular by their services like increased broadband and its connectivity and gained about 25.5% CAGR. This led to the rise of RFID Generation in retails and more enterprises in offering inventory and deliver chain integration. The IoT gadgets started coming up with increased features, with the options of assemblies, stocks, and other apps. So, the production and maintenance of IoT gadgets increase by continuously monitoring their functionalities and their impacts. More enterprises started gaining by installing RFID technology, resulting in increased Returns on Investments (ROI) with fewer labor charges and fewer funds.

6.3.3 By Component

The Industry-oriented IoT led the market with increased usage of gadgets and its implementation in areas like retails, agriculture, manufacture, automotive and logistics, and so on. This increased use of IoT gadgets created a massive demand for hardware parts in the market. The manufacturing of gadgets requires hardware parts like sensors, actuators, semiconductors, microcontrollers, and RFI. These parts are the backbone for the manufacture of IoT Devices. More advancements such as Information Search and Liability management are incorporated in IoT devices via the abovementioned hardware parts.

Moreover, software components became highly demanding due to massive growth in Structuring Data Repositories, enhancing security factors, and future implementations due to Cloud-based applications' innovation. The industrial IoT initialized integrating IoT Gadgets like personal Laptops, Computers, Mobile Phones, Tablets, and so on, to limit the hardware parts manufacturing and increase its operational performance by software assistance. This idea succeeded and lead to increased market growth.

Many enterprises started their IoT Platforms with their solutions to include professional services such as Requirement Consultation, Educational needs, Environment Installations by their own manufacturing. Moreover, they offer additional services such as increased security, hardware and software maintenance, and call centers [11]. These services are employed through enterprise representatives, outside dealers, and IoT service vendors via certified employees and trained personals. IBM is a famous enterprise that offers direct control on organization and center management via expert people and trained professionals.

6.3.4 By Application

From 2019 to 2025, CAGR increased its manufacturing to 23% by incorporating more hardware components. The CAGR solutions are adopted to monitor the operators continuously and ensure operational maintenance via intelligent manufacturing solutions. The LTE-IoT offers additional services like long-time connectivity, manufacturing control, continuous monitoring, and gadgets connectivity [12]. More enterprises started focusing on reduced production costs and increased operations with a slight change in manufacturing design of Industry-oriented IoT with multiplied operations. The highest demand for IoT gadgets with increased technical design structure in all enterprises leads to a decrease in product problems and manufacturing errors in the hardware manufacturing companies. The industry-oriented IoT provides a solution for managing excessive controls and management using various technological improvements and components with existing manufacturing architecture by reducing labor needs and manufacturing prices.

6.3.5 By Region

Many Industries like machinery, automotive, and beverage manufacturing sectors in European countries demanded the IoT architecture highly due to its additional benefits to target users. However, European people faced an increase in labors and a hike in employee salary for installing digital connections and IoT applications to the users. So, more support is given by its neighboring countries in setting up IoT architecture and Liability frameworks to help people with new Improved IoT features, which resulted in market growth and creating different opportunities for the people. So, the European government brought up an initiative, namely, EPI (European Platform Initiative—IoT), to implement industry-oriented IoT features to many enterprises located within their borders. This EPI signed seven IoT-based projects, namely, symbloTe, blooper, Big0IoT, AGILE-IoT, Vicinity, TagltSmart, and Inter-IoT year 2020, to benefit the users. For example, the BIG-IoT project benefited many retailers and enterprises through different services like Continuous connectivity, Monitoring Users, free parking spots, and so on.

6.4 Industry Digitization with IoT

As the Advancement 3G Networks, industry-oriented IoT took the help of 4G connections to extend CAT-M/NB-IoT and Broadband LTE-IoT services. The newly invented IoT applications were designed in such a way that they can support CAT-M/NB-IoT communications based on 5G Networking solutions. The broadband-IoT supports a wide variety of networking gadgets installed with 5G radio and intermediate communications too. This, in turn, 5G Networks started supporting time-critical IoT devices for their time-series communications. 3GPP brought up a standard in integrating 5G networks with Ethernet applications, mainly focusing on business IoT, which benefits extra features. The Market rise in Industries with IoT services is connected automatically.

While building 5G Networks, 4 IoT connecting modules were deployed for in-person use or by gadget number to connect in a network. Many of the gadgets are built with the support of more than one IoT connecting module for providing many services. For example, a four-wheeler was built with an IoT gadget with many services like GSM, Cloud Technology, Voice over communications, and so on. Figure 6.2 depicts the IoT categorization based on industrial growth.

6.4.1 Massive IoT

Due to technological advancements, there is a need for manufacturing gadgets with less energy, reduced price, and comprehensive network connectivity. Therefore, massive IoT was invented to provide reduced latency and significantly lower cost output mechanisms [13]. Massive IoT resulted in massive demand for providing extensive area communication, comprehensive area connectivity, increased communications, and a brand for IoT Modules. Earlier, target users felt the disadvantage of low battery gadgets. Massive IoT brought advancements in computations, connections, sensor technology, electrical and electronic field, and Manufacture battery industries. So, now low-battery gadgets are replaced with actual sensors, actuators, and nodes with huge battery strands by integrating them with cloud systems, which monitor the management of user's applications. The cloud-based IoT gadgets result

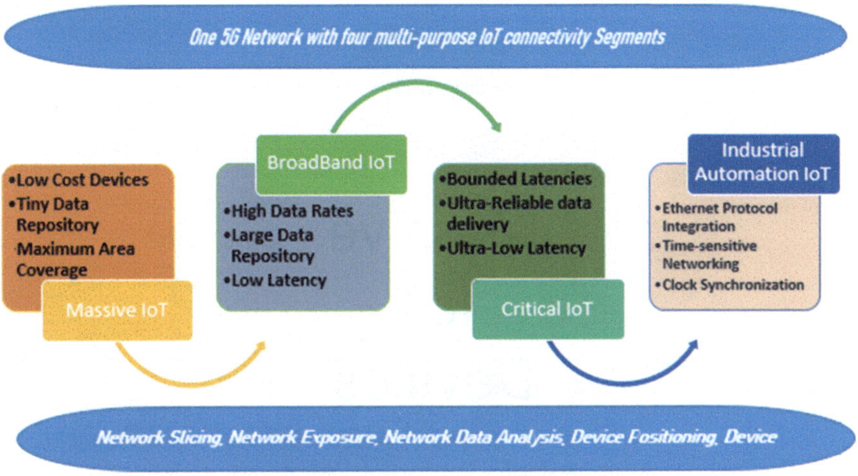

Fig. 6.2 IoT Categorization based on industrial growth

in modifying software architectures to adapt 5G Networks with more extensive IoT software. Massive IoT made changes in many software domains like health-monitoring applications, logistics management, intelligent city monitoring, home management, Smart architectural building, intelligent manufacturing, and so on. The significant advantages raised due to these factors are reduced installation price.

1. Less components required.
2. Built-in long-battery gadgets.
3. Large span connections.
4. Improved education contents.
5. New improved chatbox with safety issues.

Massive IoT gadgets majorly faced two challenges. One is covering a larger span of n communication devices in a cost-effective manner. The second is to monitor and manage all the gadgets continuously. Most of the Massive IoT software has undergone security and faith factors while exhibiting communication between point-to-point devices and different devices. This problem was rectified by employing KNN (K-Means Neighborhood Node) algorithm for enabling communication between IoT-enabled devices. Also, they gained statistics reports for instance during their point of communication. Figure 6.3 depicts the relationship among the five expertise domains, which are also called critical factors of Massive IoT Gadgets. They are communication protocols, connection parameters, security factors, identification solutions, and smart-grid gadgets. These key factors contribute to the development of massive IoT devices and supports manufacturing IoT gadgets with the integration of multiple IoT modules.

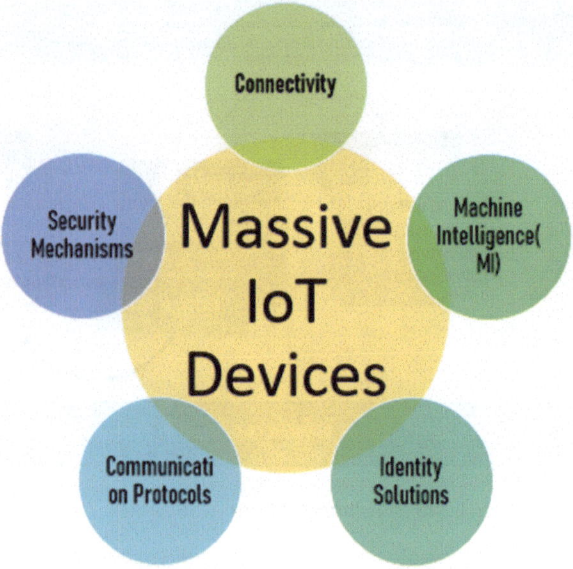

Fig. 6.3 Implementing massive IoT devices on various domains

6.4.2 Broadband IoT

Broadband IoT connectivity presents an advanced performance in lower latency and higher throughput than the extensive IoT section. Standard packages are superior wearables, aerial and ground motors, and AR/VR-enabled gadgets and sensors that require more skills than CAT-M or NB-IoT can offer. LTE has a variety of tool categories well-suited for such packages. For instance, LTE is already providing cellular connectivity to tens of millions of modern-day vehicles. There are LTE-capable clever watches inside the marketplace, and LTE-related drones are coming next. LTE gives excessive spectral efficiency and data fees, low latencies, and enhanced device battery existence and improved coverage. With advanced multi-antenna solutions and service aggregation, LTE enables total charges inside the multi-Gbps variety. Delivered to this, there are mechanisms for fast connection established order and statistics delivery. With immediate transmission schemes, the radio interface latency can be as low as about 10 ms. The vendor particular LTE scheduler has superior priority dealing with mechanisms to offer advanced performance to customers' particular institution.

LTE devices can operate in an extended coverage mode even as assisting low-throughput IoT programs. An optimized switching mechanism among regular coverage and prolonged insurance modes allows a consumer to reveal in superior records charges while determining exact insurance and entry to low information costs in difficult radio situations. For instance, a connected car can guide high facts rate packages, including infotainment, in an area with moderate coverage; however, while it has miles parked underground in difficult radio situations, it can utilize the coverage extension mode to enable low information price applications. The features for more suitable battery life (eDRX and PSM) are also supported. They can be beneficial for a few packages, as an example, for intelligent wearables and tracking of vehicles. The creation of NR goes to make bigger the capability of Broadband IoT. NR-based Broadband IoT could function in both vintage and new spectrum with tons wider bandwidths and new functionalities to guide even greater throughputs, extending into the tens of Gbps and lowering latency to approximately 5 ms. Figure 6.4 shows the Broadband IoT Services integrated with Cloud Platform.

6.4.3 Critical IoT

Critical Internet of Things (IoT) connectivity is ideal for an extensive range of time-critical use instances across most industry verticals. Cellular network operators are uniquely positioned to deliver it. Cellular Internet of Things (IoT) uses transformation throughout several sectors by enabling modern offerings for consumers and enterprises. There are presently more than one billion cell IoT connections, and Ericsson forecasts that there will be around five billion connections with the aid of 2021. As 5G deployments acquire momentum globally, corporations in nearly every industry are exploring the manageable of 5G to change their products, services, and businesses radically. Since the necessities for wi-fi connectivity in unique industries vary, it helps crew them into four excellent IoT connectivity segments: Massive IoT, Broadband IoT, Critical IoT, and Industrial Automation IoT [14].

While massive IoT and Broadband IoT exist in 4G networks, critical IoT will be added with different advanced 5G networks. Commercial Automation IoT, the

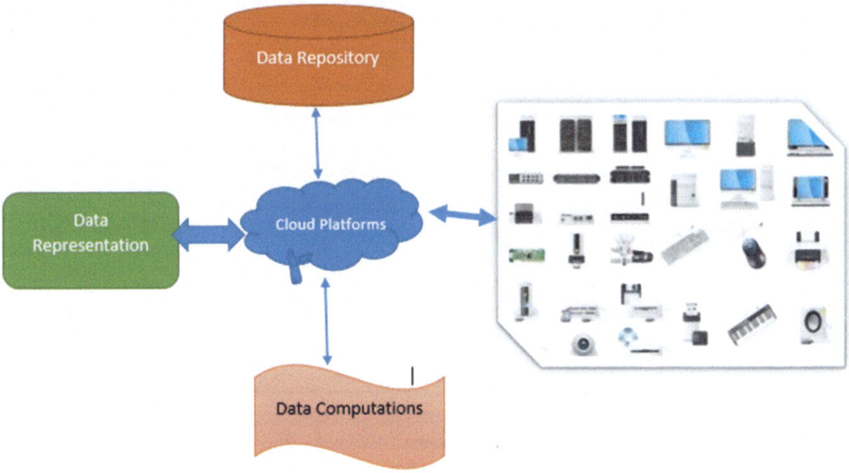

Fig. 6.4 Broadband IoT services integrated with cloud platforms

fourth phase, consists of capabilities on top of critical IoT that permit integration of the 5G machine with real-time Ethernet and time-touchy networking (TSN) utilized in wired industrial automation networks. Vital IoT addresses the time-essential verbal exchange needs of individuals, businesses, and public institutions. It has far been supposed for time-vital programs that demand records delivery within a distinctive period with required assure (reliability) tiers and records delivery inside 50 ms with 99% chance (reliability). Vital IoT is a paradigm shift from the enhanced mobile broadband (eMBB) connectivity. The information rate is maximized with no guarantee of latency. Many enterprise sectors have already started out piloting time-crucial use instances. Figure 6.5 shows the different cases enabled by critical IoT.

6.4.4 Industrial IoT

The "industrial" in industrial IoT is a little misleading. Some may say the time period applies to IoT applications used in a manufacturing setting; manufacturing is, in fact, estimated to be the biggest IoT platform by 2021, accomplishing a market fee of $438 million. It is also main the pack in terms of enterprise investment in IoT. But in my view, industrial IoT encompasses system connectivity throughout all industries (not only manufacturing):

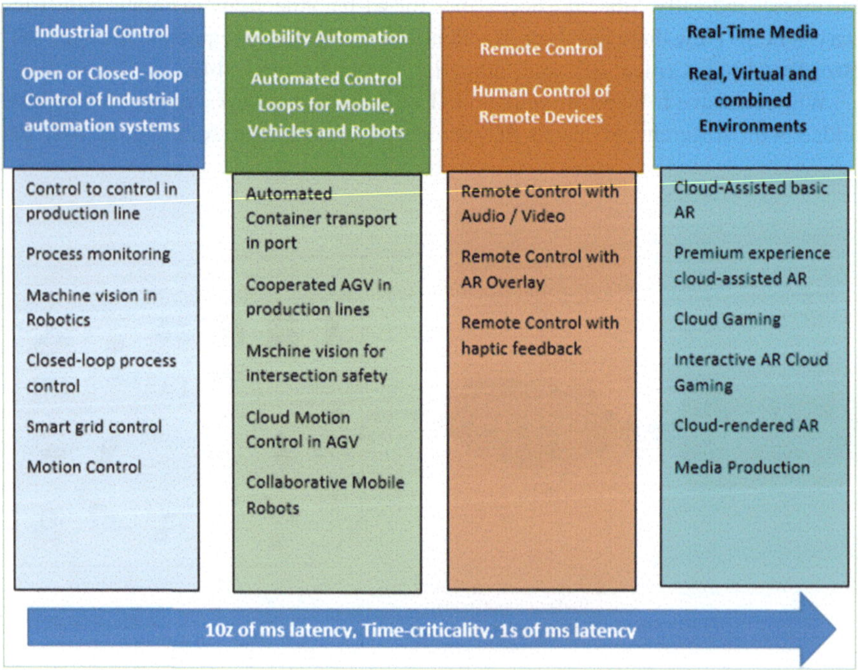

Fig. 6.5 Cases enabled by critical IoT

- In health care, wearables, mobile apps, and statistics-primarily based diagnostics may want to revolutionize the way patients access and get hold of medical services.
- In enterprises, companies can take advantage of smart homes to optimize power use. (Almost 1/2 of the energy used by homes is wasted.)
- In retail, business IoT applications can increase delivery chain efficiencies.
- In transportation, related sensors can assist with cold chain control, gasoline consumption optimization, and fleet protection.
- Utilities are already deploying smart grid technology which could help come across and isolate outages, take care of precedence routing to emergency offerings, and combine client-owned strength-era systems into the grid.

While the industrial IoT has a lot of manageable for producing price savings and efficiencies, it poses a practical task in terms of implementation. Most businesses have already invested an enormous quantity of money in operational technological know-how (OT) that was not necessarily supposed to be connected beyond the nearby network. Although there are trendy ways of speaking to instruments on buses (Modbus and Profibus, for example), there is no widespread way of connecting to the internet. How will the data movement be captured and routed again into existing information administration software program systems, whilst making positive it's all secure? The actual innovation in the years to come will be figuring out how to take it from the local, linked bus of contraptions in a production facility and get facts out from the cloud for the purposes of analytics.

For example, say you make instrumentation for meal producers that measures the temperature of the ingredients and the number of additives going into the stream. That instrumentation is useful, but it creates a closed-loop process. Figure 6.6 depicts the industrial growth in IoT [15].

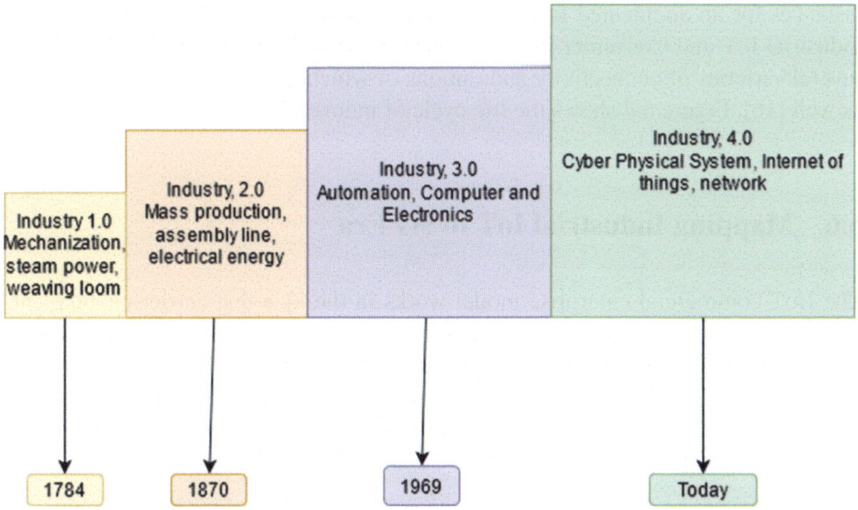

Fig. 6.6 Growth in industrial IoT

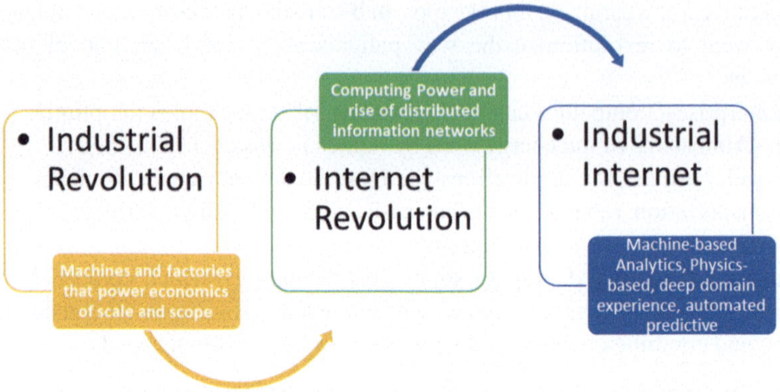

Fig. 6.7 Lifecycle of industrial IoT

6.5 Industrial IOT in a Context of Industrial Transformation

Industrial IoT in the in the past noted feel was once commonly used to make a difference between the use cases, genuine utilization, and unique applied sciences as leveraged for initially ordinarily manufacturing (smart factory) and, later, other industries on one hand and company IoT and patron IoT applications on the other.

This distinction, of course, is incredibly synthetic, and on all stages, there are overlaps. The fastest developing categories of IoT use cases, for instance, are cross-industry. Moreover, although some technologies, architectural frameworks, and functions throughout all IoT layers differ (edge computing and fog computing are every day in Industrial IoT, there are only kinds of network and connectivity tools, IIoT gateways serve different purposes, Industrial IoT platforms aid other use cases than IoT systems overall, digital twins usually are about industrial markets, the use instances for an augmented fact are no longer the identical and so forth) between Industrial IoT and Consumer IoT, an average massive IIoT challenge will leverage several varieties of connectivity and options of which some are used in patron IoT as well [16]. Figure 6.7 shows the life cycle of industrial IoT.

6.6 Mapping Industrial IoT to 5G Era

The IIoT commercial enterprise model works in the idea that carrier enablement, cloud, protection, and connectivity perform as a mixed infrastructure. IIoT imaginatively unites communications and automation—harnessing facts analytics, system-to-gadget (M2M) verbal exchange, and manner automation to enhance typical efficiency, launching a new technology of market opportunities and economic increase. Smart factories that utilize IIoT are an essential part of today's smart

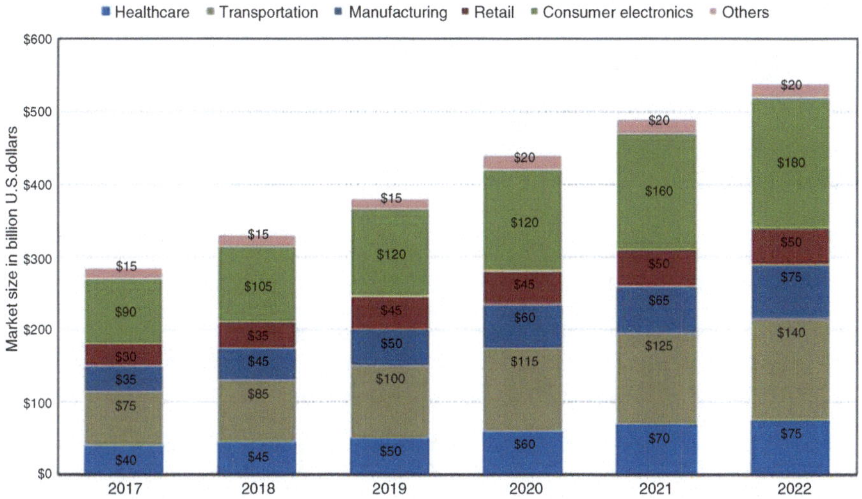

Fig. 6.8 IOT market analysis by a different domain

infrastructure, imparting an effective platform for destiny possibilities. 5G-powered technology presents a comprehensive and value-efficient answer for any production situation [17]. However, as Ulrika Engström from Ericsson illustrates, possibly the maximum massive societal development is the way IIoT puts clever gear in the fingers of intelligent humans, central to comprehensive wellness and a better-informed body of workers. By empowering human beings with the proper statistics and devices, the intelligent factory aims to revolutionize production conditions, transporting it right into a shiny, ambitious destiny. The Fig. 6.8 shows the IoT Market analysis by a different domain [18].

6.7 Conclusion

The Industrial IoT market is in its infancy. However, there is much excitement, as it is now clear that the vision can certainly be realized in sensible networks and soon. However, while large-scale Industrial IoT feasibility is now unquestioned, constructing structures' unique way is now not but clear. As for the broader IoT market, many technologies and protocols can play an essential architecture section. It will take several years to decide the first-class recipe for one-of-a-kind community types. Several disciplines are required to build an industrial IoT system. The sensors themselves ought to be miniaturized and often battery-powered; the network ought to furnish universal, seamless, and impervious connectivity; part computing will be required in many cases to system data shut to the sensor for the quick response.

Big-data techniques will need to be mixed with desktop studying—artificial talent to attain quality analytical insights. Each layer of this model is a market in itself, with competing companies and technologies.

References

1. V. Anbarasu, P. Karthikeyan and S. P. Anandaraj, "Turning Human and Food Waste into Reusable Energy in a Multilevel Apartment Using IoT," 2020 6th International Conference on Advanced Computing and Communication Systems (ICACCS), Coimbatore, India, 2020, pp. 440–444, https://doi.org/10.1109/ICACCS48705.2020.9074170.
2. Pyattaev, A.; Hosek, J.; Johnsson, K.; et al., "3GPP LTE-Assisted Wi-Fi Direct: Trial Implementation of Live D2D Technology", ETRI Journal, vol. 37, no. 5, November 2015: pp. 114.
3. Meryem Simsek; Adnan Aijaz; Mischa Dohler; Joachim Sachs; Gerhard Fettweis, "5G-Enabled Tactile Internet", IEEE Journal on Selected Areas in Communications (Volume: 34, Issue: 3, March 2016) Page(s): 460–473.
4. I.F. Akyildiz, P. Wang, S.C. Lin, SoftAir: a software defined networking architecture for 5G wireless systems, Comput. Netw. 85 (C) (2015) 118 .
5. Waleed Ejaz; Mohamed Ibnkahla, "Multi-band Spectrum Sensing and Resource Allocation for IoT in Cognitive 5G Networks", IEEE Internet of Things Journal, Year: 2017, Volume: PP, Issue: 99, Pages: 1–1
6. Andrew Girson, "IoT Has a Security Problem Will 5G Solve It?", [Available on line 15 Jan 2018], https://www.wirelessweek.com/article/2017/03/iot-has-securityproblem-will-5g-solve-it
7. J. Jin, J. Gubbi, S. Marusic and M. Palaniswami, "An Information Framework for Creating a Smart City Through Internet of Things," in IEEE Internet of Things Journal, vol. 1, no. 2, pp. 112–121, April 2014, https://doi.org/10.1109/JIOT.2013.2296516.
8. L. Vangelista, A. Zanella, and M. Zorzi, "Long-range IoT technologies: The dawn of LoRaTM," in Future Access Enablers of Ubiquitous and Intelligent Infrastructures, 2015, pp. 51–58: Springer.
9. Shilpa Talwar, Debabani Choudhu, Konstantinos Dimou, Ehsan Aryafar, Boyd Bangerter, Enabling Technologies and Architectures for 5G Wireless Kenneth Stewart Intel Corporation, Santa Clara, CA.
10. Christopher, K.K., Arul, X.V. and Karthikeyen, P., 2019, April. Smart Toll Tax Automation and Monitoring System Using Android Application. In 2019 IEEE International Conference on Intelligent Techniques in Control, Optimization and Signal Processing (INCOS) (pp. 1–6). IEEE.
11. Michael Nunez, "What Is 5G and How Will It Make My Life Better?", [Available on line 14 Jan 2018], https://gizmodo.com/what-is-5g-andhow-will-it-make-my-life-better-1760847799
12. P. Dongbaare, S. P. Chowdhury, T. O. Olwal, and A. M. Abu-Mahfouz, "Smart Energy Management System based on an Automayed Distributed Load Limiting Mechanism and Multi-Power Switching Technique," 51st International Universities' Power Engineering Conference, Sep. 2016.
13. Bridgera, "5G Promises New Horizons for IoT Solutions", [Available on line 14 Jan 2018], https://bridgera.com/5g-promises-new-horizons-foriot/
14. John Blyler, "Top 5 RF Technologies for 5G in the IoT", [Available on line 15 Jan 2018], http://www.mwrf.com/systems/top-5-rf-technologies5g-iot.
15. Cheng, Jiangfeng, Weihai Chen, Fei Tao, and Chun-Liang Lin. "Industrial IoT in 5G environment towards smart manufacturing." Journal of Industrial Information Integration 10 (2018): 10–19.

16. Schermuly, L., Schreieck, M., Wiesche, M. and Krcmar, H., 2019. Developing an industrial IoT platform–Trade-off between horizontal and vertical approaches.
17. Varga, P., Peto, J., Franko, A., Balla, D., Haja, D., Janky, F., Soos, G., Ficzere, D., Maliosz, M. and Toka, L., 2020. 5g support for industrial iot applications–challenges, solutions, and research gaps. Sensors, 20(3), p. 828.
18. P. Karthikeyyan, S. Velliangiri and M. I. T. Joseph. S, "Review of Blockchain based IoT application and its security issues," 2019 2nd International Conference on Intelligent Computing, Instrumentation and Control Technologies (ICICICT), Kannur, India, 2019, pp. 6–11, https://doi.org/10.1109/ICICICT46008.2019.8993124.

Chapter 7
IoT-Based Ensemble Method on PCG Signal Classification to Predict Heart Diseases

Esther Daniel, S. Durga, S. Iwin Thanakumar Joseph, D. Angelin, and S. Benson Edwin Raj

7.1 Introduction

Recent years have an uprising trend on IoT for healthcare applications. As diseases are rapidly spreading every day, efforts to detect the illness and its cure have to be considered. Health care plays a major part in life. Unfortunately, the growth of the population and the increase in chronic illnesses have played a major role in the healthcare system. We aim to solve healthcare problems by showing excellent care in an emergency of patients. Under these circumstances, IoT provides a solution for remote health monitoring of patients from the hospital and home. Cardiac disease is a leading medical issue and the major cause of death globally. A diagnosis could become faster and inexpensive in case of identifying the issues in the initial stage. Valve-related heart disease could be diagnosed accurately and quickly with a stethoscope. It needs good experience and listening skills to examine heart sound. However, this listening of heart sound remains one of the inexpensive methods for heart monitoring since its basic apparatus. So, the stethoscope is used for the analysis of heart disorders in all medical services. In those medical service centers in an emergency, they do not have much expertise in the diagnosis of heart disease [1].

E. Daniel (✉) · S. Iwin Thanakumar Joseph · D. Angelin
Department of Computer Science and Engineering, Karunya Institute of Technology and Sciences, Coimbatore, India
e-mail: iwinjoseph@karunya.edu; angelind18@karunya.edu.in

S. Durga
Department of Information Technology, Sri Krishna College of Engineering and Technology, Coimbatore, India

S. B. E. Raj
Department of Computer Information Sciences, Higher Colleges of Technology, Fujairah, United Arab Emirates
e-mail: braj@hct.ac.ae

S. Velliangiri et al. (eds.), *Secure Communication for 5G and IoT Networks*, EAI/Springer Innovations in Communication and Computing,
https://doi.org/10.1007/978-3-030-79766-9_7

Hence, a digital stethoscope will help to analyze easily for further treatment in an emergency.

The digital stethoscope is a device used to capture the blood flow function and heart cycle. These sounds show the heartbeat, rhythm, sound-like, importance of the aortic valve, and supply useful medical expertise. They even provide medical details about the importance of the flow of blood circulation in cardiac valve function. The acoustical stethoscope is used to capture heartbeat and other sounds for medical analysis in valvular disorder in the heart. The graphical representation of the heartbeat is called PCG signals. For examination of cardiac, PCG is very essential. When the heartbeat becomes abnormal, common symptoms like cardiac disorder and other valvular disorders occur. Thus, examination of the cardiac cycle using a stethoscope is the main test performed by doctors to evaluate the heart condition.

There was research about phonocardiograms or heartbeat a long time ago. The PCG signal is an essential part of heart treatment because it is a noninvasive, cost-effective, and efficient diagnosis for cardiovascular disorder and can be replaced without any risk to the case [2]. Cardiac analysis through an acoustic stethoscope also needs good listening skills and expertise. Higher costs for heart treatment and modern lifestyle cause major death in young age [3].

Medical innovations also enabled work through both the development of new domains and the improvement in established digital data analysis techniques. This approach in technology toward this medical issue is acceptable because the principle in the scientific analysis was much ahead of innovation used during medical conditions. The basic aspect discussed by this study seems to be noise, acoustic sound that the cardiovascular generates, their retrieval from the PCG audio database, and classification of cardiovascular diseases. Abnormal symptoms of the cardiac-specific sounds are distinct from those of the "ordinary" heart sound [3]. Therefore, modulation of such heart sound can be used to identify and diagnose cardiovascular disease. Earlier efforts [4, 5] have been undertaken in the field of digital signal processing for a heartbeat which provided the history for this study.

The typical clinical procedure is to lay down the patient in the supine position and perform silent respiration in a quiet clinical setting. The digital or electronic stethoscope is placed in the mitral area. The digitally processed waveform records the phonocardiograph feature, with time intervals in ms. Thus, the systole and diastole s1–s2 heart sound and electromechanical systole are acquired. Those are necessary to analyze features of the cardiac, valves, and also to identify cardiovascular diseases. The heart cycle has two sounds: the first is s1 and the second is s2. At diastolic contraction time, the s1 occurs in a particular time consists of the aortic valve closing, pulmonary artery, and aortic valve opening. Whenever the ventricle stress increases the arterial pressure, the mitral valve opens, the blood vessel pressure decreases below the systolic blood, and the mitral valve closes, triggering s2. The efficiency decreases sharply as in the pulmonary artery, eventually leading to s3. In the same way, the aortic systole can also form a loud s4 rhythm of the heart [6]. These heart sounds (S1, S2, S3, and S4) carry clinical information and are used in clinical diagnosis. Murmurs are noise-like sounds and defects in heart valve functioning caused by heart diseases. PCG analysis helps to diagnose different forms of cardiovascular disease, in particular, mitral valve disease, and extremely accurate cardiac rhythm, pressure including heart valve pumping efficiency.

The remaining part of the article is arranged as follows. Some related studies are reported in Sect. 7.2. In Sect. 7.3, classification of heart sound using various feature extraction block diagrams about the database used for this work and heart segmentation. Section 7.4 emphasizes feature extraction and classification algorithms. Section 7.5 describes the results of the experiment and evaluations. Section 7.6 concludes the paper.

7.2 Related Work

The classification of methods for cardiac sound usually consists of three main phases: The first step focuses on the identification and segmentation of conditions such as S1 and S2 to distinguish individual heart cycles; the second step relates to the last phase extracting features that provide the input: classification of a feature. At last, the classifier algorithm is used to classify.

The description of the Hs1 and Hs2 cardiac sound is taken from the specific phonocardiogram signal. The healthy and unhealthy signal for treatment of the heart is classified by using this system. The methodology of the classifier achieves 91% accuracy. The healthy and unhealthy signal for treatment of the heart is classified by using this system. Bayesian classifier gets better precision [7]. The heart sound characteristics were extracted using linear band frequency cepstral of the heart sounds. The proposed method is a wavelet-based algorithm. The identification of heartbeat involved the method of distance similarity [8]. A new approach was proposed based on cepstral features for the wavelet packet. The presented functions were using the banks of the nonlinear wavelet filter packets. They are matching the rhythm of a heartbeat to the acoustic design. Using the HSCT-11 database for evaluation of the proposed program, from the results obtained, the best accuracy through identifying the system proposed was 91.05% [9]. The technique of classification is based on a designed spectrogram and decomposition with tensor. The frequency components of the cardiac cycles that were identified have been measured to a specified length. The scaled spectrograms extracted the most discriminative features through the dimension reduction method. Using the tensor decomposition method, the characteristics of the scaled spectrograms were extracted. The proposed algorithm was tested on the data sets sponsored by the PhysioNet challenge in 2016 and the heart sound challenge classification PASCAL. The maximum normal accuracy was 96%. Several of the academic literature up to now have used varying PCG databases to validate methods for the classification of the heartbeat. A major problem in computer-aided cardiac diagnosis is the insufficient number of open PCG data sets for widespread clinical validations. Similarly, with the support of the Physionet Challenge PCG open dataset, medical validations were made available to the research world [1]. The segmented PCG signals are classified into spectral amplitude and wavelet entropy. They presented two characteristics and trained mostly on features extracted from a decision tree model. They noted an 80% Ac for PCG signals with low noise [10]. The related features learned from the study of time-frequency, wavelet, and statistics. They used a classifier ensemble to improve the performance of the cardiac sound detection system, and they mentioned an overall accuracy of 80%, which also needs previously segmented PCG records [11]. These classifications of PCG signals introduced matrix norm sparse coding and combined it with time-domain functions. They merged the sparse coefficients of coding with many other time-domain features and used them to train a collection of the classification method. They registered a balanced precision of 89.26% [12]. Split the PCG signal into various frequency ranges from low to ultrahigh levels. For such frequency range, they extracted amplitude envelopes which summed up to an average of 53 characteristics. They developed a technique of possibility in evaluation to execute the selection and classification of features. They mentioned a test score of 85.3% for Ac in the challenge dataset [13]. Mel-frequency spectral coefficients (MFSC) and a deep convolutional neural network (CNN) are used to classify PCG signals. MFSC developed a three-dimensional image array, which was used to train CNN. For training, they used 520 normal images and 567 anomalous images. They mentioned the best score for the hidden test dataset of 86.02% [14].

7.3 Proposed Heartbeat Classification Method

In the proposed method shown in Fig. 7.1 the classification methods includes certain steps in the process like segmentation of heart sound, Identification of s1and s2, and denoising. Feature extraction is done. The novel approach on temporal feature

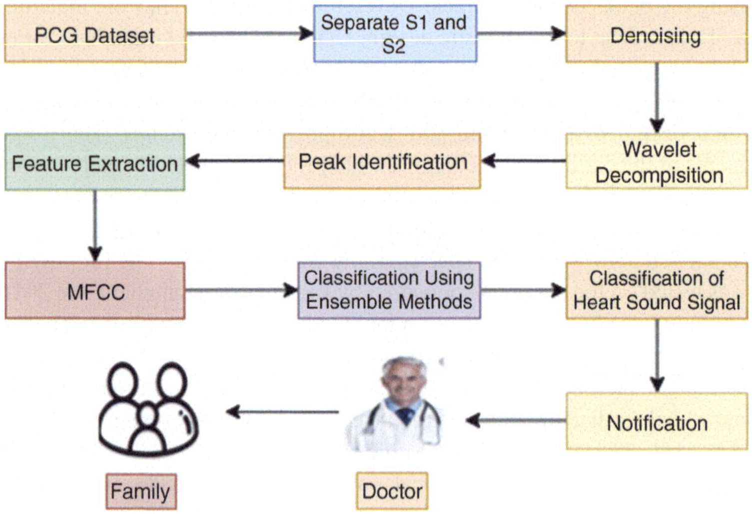

Fig. 7.1 Proposed method for classification of heart sound signal

and MFCC feature to extract features from heart sound. The proposed methods aim to identify the heart sound as regular or irregular from the dataset. For this, the Classifier algorithms are used. The best result achieves using the Ensemble method with better accuracy, sensitivity, specificity, and running time.

7.3.1 Dataset

This study is based on data collection Physionet challenge 2016 training and validation of heart sound system classification. The dataset includes PCG reports of patients with both stable and heart diseases. The data contain heart problems including valves block, MR, and pulmonic valves. These are generally called abnormalities of heartbeat in phonocardiogram reports. In the dataset document, usable numbered a to f. The repository has become a set of documents across various healthcare organizations that have specific data systems. The time limit of each recording is less than 2 m [15]. Detailed information on the 2016 challenge dataset from Physionet is extracted for implementation.

7.3.2 Separation of s1 and s2

The separation of s1 and s2 was a topic of several surveys. This represents the first move in the study of heart tone and perhaps the most challenging. The "envelope analysis" approach can be considered the most common approach to segmentation. This method measures the envelope signal as cardiac sound, identifies signal envelope peaks, decides that range are s1, s2 afterward the heart cycle interval between s1 and s2. The horizontal line marks the threshold. The high peak is identified by the signal portion among two sequential crossing threshold levels and their position of a segment with utmost analysis. Then analyses grouped each with the past three peak positions. The gap between the first and the second peak is estimated and contrasted with the difference between the second and the third peak time. The systolic is less duration than the diastolic; this means that s1 has to be the top left of systole, or equivalently s2 has to be right. These identifications have another high point provides for all other peak positions. The S1–S2 intervals then form boundaries of the cardiac cycles, and the segmentation is completed [16].

7.3.3 Denoising

The Discrete wavelet transform technique is utilized mostly on PCG signals to determine the heartbeat. To accurately classify s1 and s2, it must find its important features, such as time and frequency domains. The longest period among two consecutive peaks was its diastolic duration extending from its later part of the s2 to the early part of s1. It displays the estimated sixth features and the description of the elements from d1 to d6. The extraction is achieved with discrete wavelet transform (db6) in the sixth level [17]. A frequency band of the power of S2 has been used to accurately identify S2. The denoised phonocardiogram signal has been analyzed by feature six until extraction. A few other abnormalities have frequencies higher than the regular noises up to 600 Hz. Because certain frequencies are always less than the 2000 Hz sampling range, important heart sound signals will not be avoided. A fifth stage discrete wavelet decomposition of the opposed signal is performed after down sampling to extract its frequencies of all other decomposed modules. After this, the reconstruction of coefficients takes place [18].

7.3.4 Peak Identification

The positions of s1 and s2 are decided once all the peaks have been identified. The s1 and s2 positions are estimated using the following statistics:

1. In particular, the diastole duration s2–s1 is greater than systole duration s1–s2 and
2. The systole duration is more constant that is s1–s2.

In identifying the position of s1 and s2, the mean value and SD of both the time interval duration of peaks are known. The long duration is measured between the two peaks. The above duration has been immediately the ventricular systole period, as well as its positions, are respectively noted as s2 and s1. The rest of the peaks are labeled as s1 and s2 using this duration as just a reference [17].

7.3.5 Extracting Feature from PCG Signal

Numerous studies are used to pick the right features because classification method output differs directly from the selection of appropriate features. The feature is extracted from the preprocessing signals. These are measured with time domain, frequency domain, cepstral coefficients domain, high order, and entropy properties. Some research was carried out to predict anomaly classification using features based on segmented PCG signals. The novel method in the extraction of features consisting of using MFCC coefficients and segmented heart sounds is new for classification with PCG. The details about features are shown in Table 7.1. The PCG monitoring envelope allows predicting both the heart cycles and other associated information [19].

Therefore, its output is a specific diagnosis of the cardiac anomaly. Time-domain functions were derived from the resulting PCG signal envelogram. After measuring

Table 7.1 Feature domains and its meaning

Feature domain	No. of feature	Physical meaning
Time	1–20	SD, mean value of a cardiac cycle. These are called the envelope of the signal
Frequency	21–40	Dominant frequency value, dominant frequency magnitude, dominant frequency ratio These are called the dominant frequency of the signal
High-order statistics	41–60	To evaluate the kurtosis of each PCG signal and its skewness
Entropy	61–80	The spectral entropy, sample entropy is used to calculate the complexity of the PCG signal
Cepstral's coefficient	81–100	The MFCC coefficients are used for the audio signals. 1–13 MFCC coefficients used

the power spectral of the processed PCG signals, three frequency-domain functions have been extracted using Fourier transform with a cutoff frequency. The skewness is used to calculate the asymmetric in normalization. It is represented by positive and negative values. The kurtosis coefficient is used to calculate features from preprocessed phonocardiogram signal. To calculate the complexity in PCG, signal entropy and spectral entropy are used [19].

7.4 Mel-Frequency Cepstral Coefficients

The MFCC represents spectral acoustic information. To extract features from the audio signal the MFCC is used, and it represents in linear. It defines frequency in a logarithmic manner, influenced by human ear behavior. It is an efficient algorithm for signal analysis that is commonly used for sound recognition fields. The benefit of obtaining MFCC variables would be that all sound signal characteristics were also focused during the first coefficients, thereby enabling the extraction task for classification techniques and also for sound identification functions [20]. The FFT method is performed for the spectral analysis of the window sliding. The transformed amplitudes are then converted to Mel scale through discrete cosine transform [21].

For every sound frame, each coefficient does have quality. The adjustments around the spectrum of different PCG signals within each coefficient are discussed. The range of a signal source from cardiac sounds and whispers could be assumed at evenly spectral peaks as regards frequency coefficient definitions to rhythmic sounds [22]. MFCC feature extraction process is shown in Fig. 7.2. MFCC algorithm is described as follows:

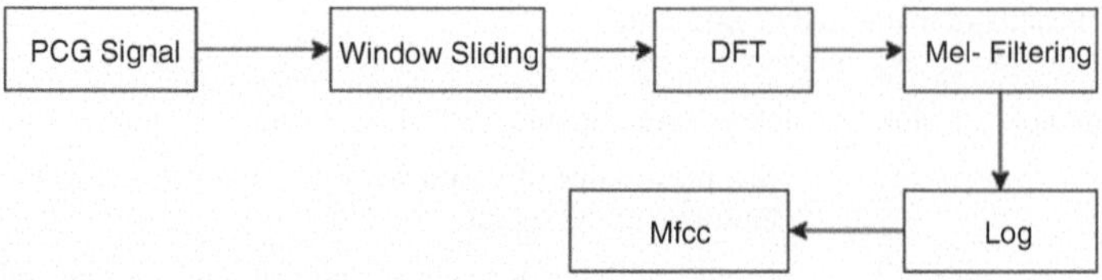

Fig. 7.2 MFCC feature extraction process

7.4.1 *MFCC Feature Extraction Algorithm*

1. The PCG signals are divides into frames. $\left(\frac{\pi n}{M}\left(m-\frac{1}{2}\right)\right)$.
2. For each frame get the amplitude spectrum by using window sliding.
3. Take the logarithm for those spectra.
4. Apply those above resultant spectrums into Mel Scale as shown in Eq. 7.1.
5. Convert to discrete cosine transform for obtaining our resultant coefficients.

The Mel scale is calculated as shown in equation:

$$M = 1127\log_e\left(1+\frac{f}{700}\right) \tag{7.1}$$

where f is the linear frequency and m is the corresponding frequency in the Mel-scale. MFCC is measured from the DCT spectrum as shown in the below equation.

$$C_n = \sum_{m}^{N=1} S_m \cos\left(\frac{\pi n}{M} m - \frac{1}{2}\right) \tag{7.2}$$

where C_n represent the nth coefficient of the cepstral, S_m denotes the result of the filter bank, and M denotes no. of channels like $n = 1,2,3\ldots.M$.

7.4.2 *Support Vector Machine (SVM)*

Support vector machine (SVM) predicted the class by constructing a hyperplane in high-dimensional space. This ML algorithm is mostly used for binary classifiers. The hyperplane is labeled as the appropriate higher dimensional space, with the maxim border from classes. This algorithm uses predicting two classes following a hyperplane generation that depends on the nonlinear dataset. The support vector machine has been described as the linear classifier, while the nonlinear classifier is represented by kernels such as polynomial, radial basis function, Gaussian radial base, and sigmoid [23].

7.4.3 K Nearest Neighbor (KNN)

K nearest neighbor (KNN) is the frequently used technique for classifier and class regression. KNN classifier output improves as the training and test samples increase. The feature vectors are estimated using Euclidean distance in the training set [24]. The class allocated the unnamed vector that affects the nearest K sample known as the voting system.

7.4.4 Ensemble Methods

The classification algorithm ensemble seems to be a set of many classifiers, whose individual decisions are combined to classify the test examples in some way. It is well known that an ensemble always performs much better than the individual classifiers that make it happen [25].

7.4.5 Bootstrap Aggregate (Bagging)

Bagging stands for Bootstrap Aggregate which one of the most efficient ensemble learning methods [25]. This generates many bootstrap training sets from the initial training set and uses to construct a classifier for addition in the ensemble. It trains with various classifiers and bootstrapped replicas of the initial training dataset. The new classifiers are trained, one for each of those new training sets. Therefore, differences have been acquired with the use of large data subsets with the resampling method. Eventually, when each classifier is given an unknown cause, a majority or weighted vote is used to predict the class [25].

7.4.6 Boosting

This model is used for combining methods to find a highly applied model to replace one another. It is akin to bagging make use of voting for our goal to classify the numeric analysis to the outcome of a single model [25]. This is similar to the Bootstrap Aggregation method except that the classification is performed within several steps. Samples, according to the classification's accuracy, are given definite weights until the final model could be created. The bagging and boosting algorithms are as follows:

Bagging Algorithm
Input

$$Dataset\ D = \{(x_{1,}\ \ y_1), (x_2, y_2)\ldots\ldots\ldots\ldots (x_n, y_n)\};$$

Base learner classifier β ;
No. of learner round S.

Process

for s =1,….,S;

$h_t = \beta\,(A,\, D_{bs})\,/\,D_{bs}$ *allocation via Bootstrap.*

end

Output

$$K(x) = \arg_{y\varepsilon\gamma} \max \sum_{t=1}^{T} \left(h_t(x) = y\right)$$

Boosting Algorithm

Input

Dataset $D = \{(x_1,\ y_1), (x_2, y_2) \ldots\ldots\ldots\ldots (x_n, y_n)\};$

Base learner classifier β ;

No. of learner round S.

Process :

$D_1 = D.$ *// Initialization allocation.*

for s =1,….,S:

$h_t = \beta\,(D_t)$; *// Train the weak allocation Learner* D_t

$\in_t = P_{x\sim D_t}\left(\left(h_t(x) \neq f(x)\right)\right)$; *// calculate (the error of* h_t

$D_{t+1} = Adjust\_Distribution\left(D_t, \in_t\right)$

End

Output

$K\,(x) =$ *Merge the output* $(\{h_1\,(x) \ldots\ldots h_t\,(x)\})$

7.4.7 Evaluation Metrics

These parameters of performance are calculated as per formulas (7.3–7.5). They are described as follows: They are discussed as follows: TP, true positive range, showing that only some unhealthy subjects are clearly described as unhealthy subjects; FN, false negative range, showing that only some participants with irregular heart sounds are recognized as healthy subjects; TN, true negative range, which means that some healthy people are correctly labeled as healthy people; and FP, a false-positive figure, which means that some healthy people are identified as having an irregular heart rhythm [26].

$$\text{Accuracy} = \frac{\text{Tp} + \text{Tn}}{\text{Tp} + \text{Fn} + \text{Tp} + \text{Fn}} \times 100\% \tag{7.3}$$

$$\text{Specificity} = \frac{\text{Tn}}{\text{Fp} + \text{Tn}} \times 100\% \tag{7.4}$$

$$\text{Sensitivity} = \frac{\text{Tp}}{\text{Tp} + \text{Fn}} \times 100\% \tag{7.5}$$

The performance measures that have been used in Physionet Challenge for 2016 and have used this to evaluate our results with those obtained and achieved in the experiment [20].

7.5 Result and Discussions

The collection of filter bank frequencies of the MFCC features is used as a hyperparameter during classification [20]. So much testing was done, with various time-domain frequency ranges, to enhance the outcome of the classifier [21]. Table 7.2 displays the outcomes achieved with the various methods, use 12 and 13 MFCC functions executed with Support vector machine, KNN, and Ensemble along with Tf. The set of measures consisting of 13 MFCC and TF showed greater results than the set of 5 and 6 MFCC and Tf. The better outcomes using Classifiers are achieved among all the classification techniques, followed by KNN and Ensemble, from this sequence as shown in Table 7.2. Their performance measures such as sensitivity, specificity, and accuracy are shown in Fig. 7.3 classification performance of the classifiers.

The outline of Table 7.2 shows that a better outcome has been achieved.

Table 7.3 displays PCG-related studies on identification and classification of heartbeat comparison based on the performance results of this study. The approaches used in this analysis include Physionet 2016 database [15].

Figure 7.3 shows the results of classification models like KNN, Support vector machines, and Ensemble. The test results for Se, Sp, and Ac were really clear while considering those classification algorithms with the three possible observations: (1) SVM accomplishes the worst outcome in accuracy compared to the other two algorithms, probably due to its too basic structure and minimal classifier efficiency. (2) KNN exhibts offers good accuracy when compared with the results of Sp and Se proving that KNN may not provide better performance due to low Sp and Se outcome [27]. (3) Ensemble gives better results with 95.3% Ac and confirms its superior classification efficiency and better Sp and Se comparing with the other two classification models.

The performance estimation is provided using a segmented PCG signal dataset. The performance is measured through training a sample of 50% and validating a

Table 7.2 Result obtained from MFCC and temporal feature

Features type	Models	SC	SP	AC
12MFCC + Tf	SVM	90	91	91.8
13MFCC + Tf	SVM	92.3	93	92.9
12MFCC + Tf	KNN	82.3	79.9	81.7
13MFCC + Tf	KNN	80.6	88.7	77.4
12MFCC + Tf	Ensemble	90.8	92	93.1
13MFCC + Tf	Ensemble	92	90	94.5

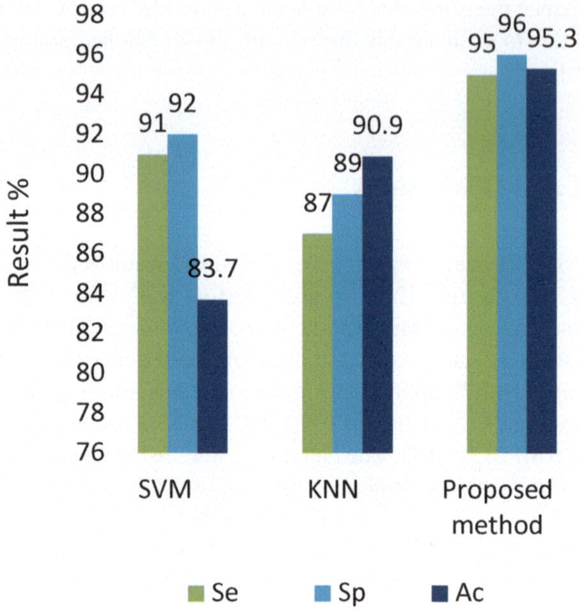

Fig. 7.3 Se, Sp, Ac, obtained from classifiers

Table 7.3 Comparative analysis of existing work based on classification of heartbeat Physionet dataset 2016

Author	Methods	Performance metrics
Nabhan Homsi [11]	Random forest algorithm, Logit boosting, Cost effective classifier	Ac: 80.13%, sc: 80%, sp: 81.45%
Whitaker, Suresha [12]	Sparse coding techniques and time-domain features	Ac: 89.5%, sc: 90.32%, sp: 88%
Plesinger, Viscor [13]	Possibility evaluation	Ac: 90%, sc: 87.8%, sp: 94.1%
Maknickas V, Maknickas A [14]	Deep Learning, CNN and mfsc	Ac: 86.02%
Langley [10]	Wavelet packet and entropy from unsegmented PCG signal	Ac: 80.6%, sc: 94.2%, sp: 65.8%
Proposed method	Feature extraction from PCG and MFCC	Sensitivity: 95% Specificity: 96% Acc: 95.3%

50% sample analysis using an ensemble approach. The evaluation analysis requires the duration of 5 s in PCG signals. Figure 7.4a, b shows the consistency of the training summary and the validation. For evaluating the performance of the model, extracted features were trained based on the supervised learning method using the ensemble classifier. The classification of an abnormal or healthy heart sound is carried out by the ensemble model. The ensemble model was trained by part of the data and validating with the remaining data.

However, various features that acquired complete information of the PCG signal based on the mfcc coefficient, power spectral density (PSD), and envelope extraction that removes complex segmentation steps are used. Our proposed ensemble-based classifier gives the best result as Ac of 95.3% with Sc of 95% and Sp of 96%.

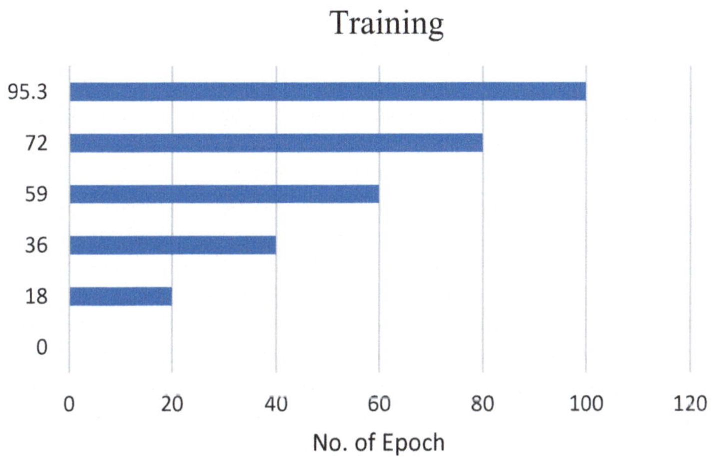

a) Time taken for Training

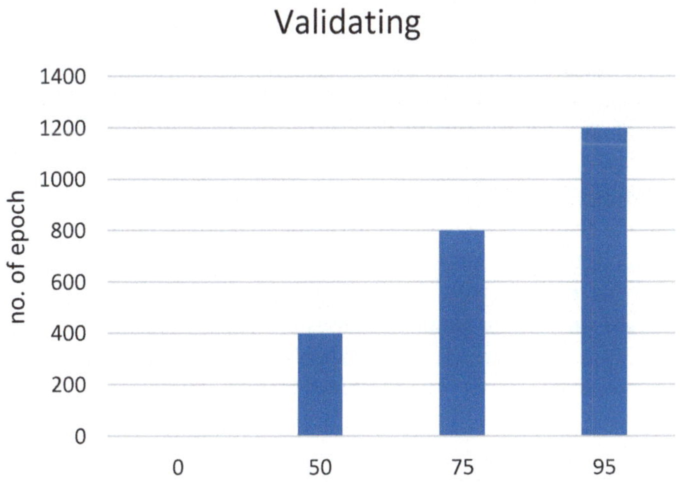

b) Time taken for Validating

Fig. 7.4 (**a**) Time taken for training. (**b**) Time taken for validating

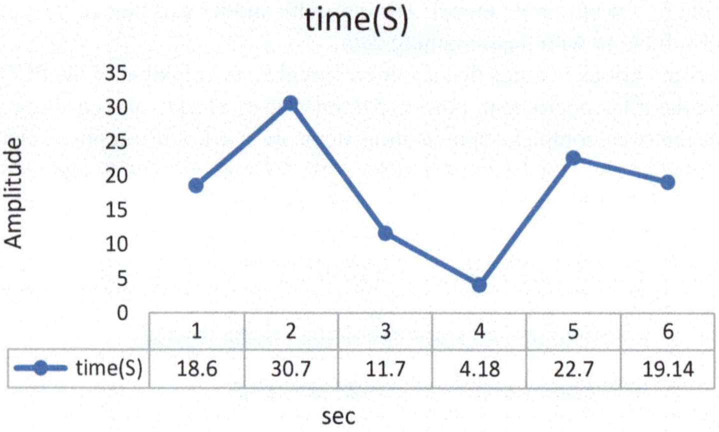

Fig. 7.5 Time requirement

Figure 7.5 shows the time spent extracting features to classifying the data. The running time of SVM is 18.6 s and for KNN is 4.18 s and then for the proposed method is 30.7 s to extracting a feature from heart sound. For that, temporal feature and Mfcc feature are used. The proposed methods give the best result in 30.7 s with better accuracy.

7.6 Conclusion

The IoT-based PCG signal classification enables to development of cost-effective and equitable healthcare innovations for managing cardiovascular disease. An efficient heart sound classification method for medical diagnosis is proposed. Initially, the system evaluates whether the PCG signal is clinically "healthy" or "unhealthy" in condition. Denoising is used to eliminate extra noise sounds and whispers. Selecting random features during training the ensemble method gives better performances with the feature. This will decrease the diagnostic duration for testing the PCG databases for the detection of cardiovascular disease, thereby allowing the doctor to provide patients with quicker medical care. The techniques for feature extraction based on various domain features are used, following on classification using the ensemble algorithm. The result shows high accuracy obtained when compared to conventional methods. The combined feature of time domain and frequency domain extracting from PCG to enhance accurate identification of cardiovascular disease is implemented. The Ensemble methods achieve better accuracy and sensitivity. For future work, the proposed model can be implemented using deep learning techniques.

References

1. Syed Z, Leed D, Curthis D, Nesta F, Levin RA, Guttag J. A Framework for the Analysis of Acoustical Cardiac Signals. IEEE Transactions on Biomedical Engineering. 2007;1(51):651–662. doi: https://doi.org/10.1109/TBME.2006.889189.
2. D.H. Bekkering, Weber J.(1994), Standardization of phonocardiography: effort in the Netherlands, American Heart Journal, Aug, 54 (2):pp. 316–322.
3. A.A Luisada, (1965) "Auscultation and phonocardiography", Mosby, St. Louis, USA.
4. H. Liang, S. Lukkarinen and I Hartimo (1997), "Heart sound segmentation algorithm based on heart sound envelogram, Computational Cardiology", Vol. 24, pp. 105–108
5. H. Ljunggren (1949), "Investigation of an apparatus for calibrated phonocardiography according to the Mannheimer-Stordal system", Acta Med Scand, Jun 30, Vol. 133 (6): pp.388–93.
6. Hassani K, Bajelani K, Navidbakhsh M, Doyle DJ, Taherian F. Heart sound segmentation based on homomorphic filtering. Perfusion 2014; 29 (4): 351–359.
7. Angelin D, Esther Daniel, Getzi Jeba Leelipuspham Paulraj, S. Durga. "A Cost Effective Framework for Heart Signal Monitoring System", 2020 6th International Conference on Advanced Computing and Communication Systems (ICACCS), 2020
8. Cheng X F, Ma Y, Liu C, et al. Research on heart sound identification technology. Sci China Inf Sci, 2012, 55: 281–292, doi: https://doi.org/10.1007/s11432-011-4456-8
9. Mohammed Abo-Zahhad, Sabah M. Ahmed, Sherif N. Abbas "Biometrics from heart sounds: Evaluation of a new approach based on wavelet packet cepstral features using HSCT-11 database" Volume 53, July 2016
10. Langley P, Murray A (2017) Heart sound classification from unsegmented phonocardiograms. Physiol Meas 38:1658–1670. https://doi.org/10.1088/1361-6579/aa724c
11. Velliangiri, S., & Premalata, J. (2020). A Novel Forgery Detection in Image Frames of the Videos Using Enhanced Convolutional Neural Network in Face Images. Computer Modeling in Engineering & Sciences, 125(2), 625–645.
12. Velliangiri, S., Karthikeyan, P., Joseph, I. T., & Kumar, S. A. (2019, December). Investigation of Deep Learning Schemes in Medical Application. In 2019 International Conference on Computational Intelligence and Knowledge Economy (ICCIKE) (pp. 87–92). IEEE.
13. Plesinger F, Viscor I, Halamek J, Jurco J, Jurak P (2017) Heart sounds analysis using probability assessment. Physiol Meas 38:1685–1700. https://doi.org/10.1088/1361-6579/aa7620
14. Maknickas V (2017) Recognition of normal–abnormal phonocardiographic signals using deep convolutional neural networks and mel-frequency spectral coefficients. Physiol Meas 38:1671–1684. https://doi.org/10.1088/1361-6579/aa7841
15. Palani Thanaraj Krishnan, Parvathavarthini Balasubramanian, Snekhalatha Umapathy. "Automated heart sound classification system from unsegmented phonocardiogram (PCG) using deep neural network", Physical and Engineering Sciences in Medicine, 2020
16. Balili, Christine C., Ma. Caryssa C. Sobrepena, and Prospero C. Naval. "Classification of heart sounds using discrete and continuous wavelet transform and random forests", 2015 3rd IAPR Asian Conference on Pattern Recognition (ACPR), 2015. Publication
17. P.S. Vikhe, S.T. Hamde, N.S. Nehe. "Wavelet Transform Based Abnormality Analysis of Heart Sound", 2009 International Conference on Advances in Computing, Control, and Telecommunication Technologies, 2009
18. Michael H. Crawford, MD (2002), "Approach to cardiac disease diagnosis", Mayo Medical School, Rochester, Minnesota, Mayo Clinic, Scottsdale, Arizona, Current Diagnosis & Treatment in Cardiology 2nd Ed, pp. 14–36.
19. Sinam Ajitkumar Singh, Takhellambam Gautam Meitei, Swanirbhar Majumder. "Short PCG classification based on deep learning", Elsevier BV, 2020
20. Diogo Marcelo Nogueira, Carlos Abreu Ferreira, Elsa Ferreira Gomes, Alípio M. Jorge. "Classifying Heart Sounds Using Images of Motifs, MFCC and Temporal Features", in Progress in Artificial Intelligence, Springer Science and Business Media LLC, 2017

21. Lubaib, P., and Muneer, K. A., The heart defect analysis based on pcg signals using pattern recognition techniques. Procedia Technol. 24:1024–1031, 2016.
22. Sinam Ajitkumar Singh, Swanirbhar Majumder, Madhusudhan Mishra. "Classification of short unsegmented heart sound based on deep learning", 2019 IEEE International Instrumentation and Measurement Technology Conference (I2MTC), 2019
23. Velliangiri, S., Sekar, R., & Anbhazhagan, P. (2020). Using MLPA for smart mushroom farm monitoring system based on IoT. International Journal of Networking and Virtual Organisations, 22(4), 334–346.
24. Leal, A., Nunes, D., Couceiro, R., Henriques, J., Carvalho, P., Quintal, I., & Teixeira, C. (2018). Noise detection in phonocardiograms by exploring similarities in spectral features. Biomedical Signal Processing and Control, 44, 154–167.
25. Emine Yaman, Abdulhamit Subasi. "Comparison of Bagging and Boosting Ensemble Machine Learning Methods for Automated EMG Signal Classification", BioMed Research International, 2019
26. Patidar, S.; Pachori, R.B. Classification of cardiac sound signals using constrained tunable-Q wavelet transform. Expert Syst. Appl. 2014, 41, 7161–7170.
27. Tien-En Chen, Shih-I Yang, Li-Ting Ho, Kun-Hsi Tsai, Yu-Hsuan Chen, Yun-Fan Chang, Ying Hui Lai, Syu-Siang Wang, Yu Tsao, Chau Chung Wu. "S1 and S2 Heart Sound Recognition using Deep Neural Networks", IEEE Transactions on Biomedical Engineering, 2016

Chapter 8
IoT-Based Model Predictive Control Architecture for HVAC Systems and a Comprehensive Study of Security Issues of 5G and IoT

S. Dhanalakshmi and M. Poongothai

8.1 Introduction

The Internet of Things (IoT) moves people from the world of secluded systems into one where devices are equipped with various sensors, are interconnected, and exchange data with each other. Heating, ventilation, and air-conditioning (HVAC) are among the industries that are to be dramatically revamped by the new IoT technologies. For manufacturers, IoT incorporates intelligence and connectivity into devices which makes them to give modernized offerings and build new business models. For clients, the initiation of the IoT brings greater degrees of comfort, better functionality, and reduces electricity bills.

Air conditioners and electric fans give about 20% of the overall electricity used in buildings around the world, according to a study by the International Energy Agency (IEA), and it is anticipated that the global energy demand from air conditioners will triple by 2050. The key benefit of IoT is to bring down energy consumption by incorporating operational visibility into HVAC appliances. By integrating sensors with smart equipment, it becomes highly possible to lower the power usage by continuously monitoring the occupancy within the building. This is achieved by turn heating or cooling down when no movements in the room are noticed through an app notification.

MPC is a higher approach used by a pair of limitations to regulate a system. The term MPC typically does not assign a specific control technique to a sufficient collection of management methods, allowing free use of a process model to get the control signal. Due to the increase in building computing capacity and the ease of use of a large volume of tracked building data, predictive monitoring of the model

S. Dhanalakshmi (✉) · M. Poongothai
Department of Electronics and Communication Engineering, Coimbatore Institute of Technology, Coimbatore, India
e-mail: dhanalakshmi.s@cit.edu.in; poongothai@cit.edu.in

S. Velliangiri et al. (eds.), *Secure Communication for 5G and IoT Networks*, EAI/Springer Innovations in Communication and Computing,
https://doi.org/10.1007/978-3-030-79766-9_8

is becoming more versatile today. For the best management of on-site renewable power sources and for successful energy storage systems, the Predictive Control model is useful. Because of its ability to understand limits, predict disruptions, and several conflicting goals in the performance of heating ventilation and air-conditioning HVAC systems, such as achieving both indoor thermal comfort and energy demand handling, MPC offers various options to improve energy performance.

Since it is a well-established technique for limited control in the field of construction and active part control, extensive attention is given to MPC. The MPC incorporates both concepts of feedback control and numerical optimization. This paves the way for the use of the capabilities of the energy luggage compartment and RES on-site output optimization. To forecast the power requirements of the generating and improve its thermal behavior on the source of the specified control aims, MPC will carry out both the forecasts of possible disruptions and the requirements given. Since MPC requires high computational demand for major optimization problems in recent decades, it has identified a sharp way to practical execution. Following the growth of modern processors, graphic processing units, and cloud storage, MPC is now commonly applied in energy systems and different types of buildings. This chapter also briefs about the 5G and IoT security challenges in the context of MPC. The section is organized as follows: Sect. 8.2 deals with the evolution of the method of the predictive control model, Sect. 8.3 shows the classifications of the MPC technique, Sect. 8.4 provides an overview of the predictive control model, Sect. 8.5 describes the use of the predictive control model in HVAC buildings, Sect. 8.6 elaborates the MPC models in building and HVAC systems, Sect. 8.7 lists the advantages and disadvantages of predictive control model in HVAC buildings, and Sect. 8.8 discusses the 5G and IoT security issues.

8.2 Historical Perspective

Some papers offering Model Predictive Heuristic Control from the late 1970s were later termed as Model Algorithmic Control, and a brief was given by some publications on Dynamic Matrix Control (DMC). To forecast the impact of upcoming control behavior on the result, the dynamic process model is used clearly in mutual algorithms. These, subject to organizational constraints, are predicted by minimizing the predicted error. In 1963, recalling the concept of the horizon, under the open-loop optimal feedback technique, which was discussed in 1970s, Propoi presented the central idea of the MPC. MPC became rapidly known in the chemical procedure sectors with the reason being the ease of the algorithm and the use of the impulse or phase response model, even if it has some drawbacks, such as having more parameters than the formulations in the conditioned space or input–output field, which is typically chosen to be more random and needs less a priori data for its recognition. The maximum number of applications on multivariable systems was carried out. Despite this progress, these principles lacked formal theories that produced results of robustness and stability, although, except in very specific cases, the finite-horizon case seems too difficult to examine.

Another important problem emerged around ideas for adaptive control, designing techniques basically articulated with input–output models for a single variable operation. This method helps to hold the anticipated performance comparable to the comparison at a time period after the delay in the phase and allows for various strategies. Extended Prediction Self Adaptive Control is the next control strategy, a constant control signal that starts from the current moment by utilizing a sub the best possible predictor instead of resolving an equation (EPSAC). In 1987, Generalized Predictive Control (GPC) was developed using the Generalized Minimum Variance principles. It is currently considered the majority common method. Multistep multivariable adaptive control, predictive functional control, unified predictive control, multi predictor receding horizon adaptive control, and so on are some of the predictive controller formulations centered on similar popular ideas.

In addition to using state-space theory, MPC generalizes itself to handle further multifaceted situations like nonlinear steps and systems of stochastic disorder and sound in the calculated variables. It is possible to handle processes with integrators by using proven state estimation techniques and expanding the phase reresponse model. The stochastic optimum control that is used for predictions without adding complications is also sensitive to implement state calculation techniques. On the source of the state observer and its stability, the MPC controller can be explained as being an adjustable one. Adjustable parameters could be used to set the poles of the observer and the regulator poles (acquired by the weights, horizons, etc) are used to assess the efficiency and power. From the Gaussian quadratic optimal linear theory's point of view, it is simple to find a detailed study of the inherent features of distinct algorithms. While the initial steps on GPC and comprehensive notes on robustness processing and the effect of philter polynomials demonstrated some basic stability theorems using state-space relationships, the main downside of this technique is the absence of original common stability outputs for finite horizon receding controls are recognized. Thus, in the 1990s, a novel series of operations on new predictive control systems with guaranteed consistency appeared. Two systems, namely SIORHC and CRHPC, were developed separately by placing endpoint equality issues on the result after a limited horizon and tested to be stable. A new approach, stable GPC, is a formulation that ensures the consistency of the closed-loop by stabilizing the steps before the objective function is reduced, taking into account all the points, and objectives specified.

With the most promising results of the stability of restricted receding horizon controls, the issue that was presumed to be too difficult ended. Although there is a solution to this problem of optimization, closed-loop stability is not guaranteed, which explicitly means there is no stability. A family of techniques was then developed using the terminal penalties and constraints, Lyapunov functions that ensure the system's stability. In the sense of state space, this regulator issue has been studied from various phases of view, and certain aid has emerged in the latest periods. Many results were obtained using robust control design aspects in the MPC background. For optimizing the objective role for the poorest state of affairs of uncertainty, the primary aim is to explicitly consider uncertainties about the MPC system and design. These results suggest that MPC will practice a still higher popularity in mutually the educational nation and the society of management practitioners. In its TDC 3000 control system, Honeywell, one of the foremost producers of dispersed control tools, utilizes robust multivariable predictive control and has announced several technological breakthroughs.

Model Predictive control is a complete linear and system technique that in the course sector is usually slow. Many complicated methods like nonlinear, hybrid, or a method with too rapid steps were exceeding the control of the Model Predictive control. In this industry, some remarkable results have been achieved over the last few years. Some restricted MPC findings have also shown that it is possible to implement a piece-wise affine controller with a little computational burden. In the literature, several applications related to hybrid and nonlinear issues also appear.

8.3 MPC Typologies

Any of the different MPC typologies can be implemented based on the design of the managed process and the form of the prediction model, which will help to solve problems with building control. Optimization algorithms can also vary depending on the existence of the optimization problem [1].

MPC problems can be either linear or nonlinear, belonging to the feature of the controlled method. The transformation of Model Predictive control from linear to nonlinear constraints is a tedious issue in reason to the efficiency of the programming of nonlinear solvers, additional computational complexity, and the absence of general-purpose nonlinear system recognition techniques.

The scale of the suspicions has impacted the system. Its influence on the model reply can be established with the specialist skill of the managed steps or during the model's data corroboration stage. When the accuracy and performance requirements for all available model variations and uncertainty magnitude within a given range are maintained, the MPC is called a robust MPC. The uncertainties are restricted in this context, and within the uncertainty spectrum, the consequential control strategy will follow the predefined constraints. An alternative resolution is obtainable by Stochastic MPC, where the possible future evolution is predicted by a stochastic dynamic process model. Random variables are initialized with the disturbances and constraints for a given probabilities distribution. In the event of continuous distributions, it becomes necessary to opt for infinite uncertainties and apply constraints within a finite probability. On the other hand, if the doubt is able to alone get worth from a small pair with a known prospect, which means if the distributions are discrete in feature, then stochastic measures can be optimized and, depending on how relevant the constraints are, constraints can be applied either in the probability or for all disruptions. For the simulation of occupancy disturbances, a stochastic approach is often used in the MPC.

All the above-described MPC strategies are formulated in such a way that it punishes if the output gets deviated from the orientation course. Usually, based on economic considerations, the upper supervisory optimization layer decides the orientation trajectory for the starting stage. By minimizing the in charge of attempt to take the course given, a Model Predictive control may improve the control sequence of the actuators at a minimum level. Another form of MPC is the economic MPC, its strategy is like where system set-points and temperature trajectories are optimized using the MPC's same cost feature.

In several energy management applications, MPC can be used where a system involves both continuous dynamics and discrete dynamics, resulting in a nonlinear discontinuity model, and such a method is called a hybrid system. View of these, the optimization is called the problem of MIP, and the building of the MPC is called the hybrid MPC. The MPC approach is commonly used in on or off, open or closed, or setting up specifications where discrete states are needed for the system process [2].

It is possible to formulate the MPC problem in two ways: one is explicit and the other is implicit. By finding the answer to the optimization constraint live for an implicit MPC, the control law is set up. The algorithm of multiparametric programming is not run online for explicit MPC to adjust the linear gain manage law as in the schedule. Wherever there is a very short computational time or the computing power is limited, the explicit MPC can be applied. This is considered as the application of explicit MPC. The types of model predictive control are discussed subsequently.

8.3.1 Classical Model Predictive Control

The first step in classical model predictive control begins with the problem definition and is followed by absolute design of the optimal unconstrained control law. The next step is the definition of settings for dual-mode prediction. Dual mode is one way of denoting it since it describes a control technique that switches between two separate laws of control. One of the ways to use the term dual mode is to break the predicted control sequences into two, one at the first *N* predicted time-steps is control inputs, and the other refers to the rest of the prediction horizon. These are called, respectively, Mode 1 and Mode 2. The classical MPC is developed based on the concepts of controlled invariance. To explain the feasibility properties of device constraints combined with terminal constraints, this is considered a convenient concept [3]. The potential trajectory measured at any given time can be extended to a subsequent moment and provides an effective method for both recurrent viability and a monotonic invariant cost property to be determined. Recurring feasibility refers to the property that feasibility is the same for instants of the current time, next time, and subsequent time. To develop closed-loop stability, monotonic invariant can be used.

8.3.2 Predictive Control Robust Model (RMPC)

RMPC mainly discusses the difficulties involved in the control issue, such as restricted uncertainty, etc. To evaluate the MPC techniques that optimize forecast through open-loop key trajectory, additive disturbances were considered in the initial process. These are distinct from approaches that use optimization over control laws that may not be known to the controller at present but would be feasible in the future after the control law is computed. This is called a closed-loop prediction strategy that takes future realizations of uncertainty into account.

The state decomposition is divided into nominal and uncertain components. In comparison to the augmented forecast state model, this definition allows for the treatment of recursive viability and robust invariance. This implies that maximum and minimum robust invariant sets are defined and calculated conveniently. An RMPC strategy can be developed accessing the dual-mode prediction paradigm and the supposed expected expenses [4] with guaranteed stability and convergence properties.

Using the game-theoretical approach, the mixed additive and multiplicative uncertainty case is discussed. This produces a complex game between the uncertainties by using the optimization of minimum–maximum, while RMPC performs the maximization. In the RMPC, by minimizing the maximum cost, the controller chooses the control input. The RMPC tube methodology is a conventional methodology, where conditions that are included are imposed by restrictions to ensure unpredictable future state and input trajectories that fall into the set sequence. This is referred to as tubes in which the future states inside the set are fully present, whose constraints on the control problem are met. The series of sets used to describe a tube is essentially referred to as polytopic sets of low complexity but are replaced by more general sets that in the future can either be fixed or scalable. Early tube RMPC strategies, and in addition, a study of an RMPC technique that achieves repeated possibility by a mock contraction of issues with respect to the full class of added uncertainty is implemented in tubes with simple poly topic cross-sections.

Open-loop methods are considered conservative since they preclude the controller from learning about potential uncertainty even if it is not available at the moment. The advantage of open-loop strategies is that it is computationally convenient. By improving the multistage min-max control issue over the entire feedback policies, the hugest pair of allowable initial circumstances and best possible output for an agreed prediction horizon and terminal control law is obtained. Its complexity is the drawback of this method, the estimation of which exponentially extends with the dimension of a horizon and device prediction.

This function can be evaluated offline in the nations of the place room specified by various pairs of active issues. Low-order systems for which and little prediction horizons, this method was not feasible, as the number of places usually grow rapidly for the size of the method. An approach to resolving this issue can be used based on the live utterance between the present state and the condition that knows best control law. During this interpolation, the lively constraint set is updated and at each active set point, the equality restricted optimization problem is solved. Although the abovementioned approach results in a significant decrease in more complex online calculations, the computationally demanding off-line computations are still needed. This will give the ideal solution a fair approximation, making it is preferable to refine a restricted set of control policies.

The method ends in a problem of convex optimization, which has a number of variables that increase quadratically with the prediction horizon by considering the previous disturbance inputs as an affine dependency in a feedback parameterization. Piece-wise affine disturbance feedback can be given in the disturbance by using separable prediction schemes with a triangular structure. The RMPC algorithm and MPC parameterized tube results show that it offers a higher volume of optimal comparable calculation load for the volume of optimization variables belonging quadratically on the horizon estimation. Complicated calculations are able to be simplified by accessing the striped triangular structure or otherwise a triangular prediction plan. The main aspect of the striped triangular structure is that it can be extended to Mode 2, and thus, it is possible for us to replace the fixed terminal law by 1. This optimization takes place in online mode, and thereby with this approach it is possible to overcome PTMPC.

RMPC can also be extended in the case of multiplicative ambiguity. The theory behind multiplicative uncertainty is based on the parameterization expressed regarding a linear state opinion law of predicted control inputs in which the optimization variable is the feedback gain that is evaluated at each moment instantly online. The technique enables that the forecast state is present in ellipsoids and uses quadratic constraints that are represented as linear matrices inequalities. For a poly-topic group of improbability, such quadratic surface assurance same, non-increasing belongings for the expected spending. Through the introduction of additional optimization different value in the mode of agitation sets useful to the expected rule, the prediction structure of this system was subsequently enriched. The major drawback to this approach is that online computing is dramatically enhanced, making it excessively demanding for systems with many undefined parameters. Via off-line optimization, this approach allows optimization of the dynamics specifying the expected state and input trajectories. This allows the maximization of the volume of the ellipsoidal attraction region [5].

The benefit of multiplicative uncertainty over additive uncertainty is that it can be treated more easily and efficiently with the implementation of tubes that describe the fruition of the forecast state and control trajectories.

8.3.3 Predictive Regulation of the Stochastic Model (SMPC)

SMPC belongs to the group of numerical optimization strategies to control stochastic systems that concentrate on constraints on the inputs of the controlled system and their states. The key feature of this strategy is using a cost function calculated along the input trajectories and ex-anticipated state [6], potential output must be quantified. These guides to a stochastic optimal control problem that is numerically brought a solution to achieve an appropriate open-loop control series or a set of criticism manage laws on the other hand. In Model Predictive Control, the optimal control problem is getting the solution by applying the first element of this appropriate sequence to the proscribed system based on updated system state details. For systems to the nonlinear dynamics and problems on inputs and states, SMPC is mainly applicable because of its numerical existence. Due to the re-repeated estimate of optimal expected trajectories [7], criticism to recompense for the impacts of ambiguity in the model is used.

Robust MPC is one type of MPC used for all model uncertainties realizations to satisfy limitations as it faces issues with tough state and input problems [8]. The key difference between RMPC and SMPC is that in most applications, RMPC is rather conservative, but SMPC offers fewer conservative solutions that operate with a broader range of constraints to be met with specific probability and average. In optimizing the predicted output to achieve it, the probability distribution of stochastic model uncertainty is taken into account [9]. This helps processes to run close to limitation limits, which is perceived to be an advantage of the MPC. Stochastic MPC is also helpful when model ambiguity is stochastic to a given dispersed probability and the restrictions are probabilistic in nature.

SMPC has a wide variety of applications in distinct areas. Some applications include the chemical and process industries, the production and distribution of energy, risk management, finance, sustainable development policy evaluation, portfolio management, climate control construction, and traffic control of telecommunications networks [10].

8.3.4 Nonlinear Model Predictive Control (NMPC)

NMPC characterized by the application of nonlinear system models for prediction is another variant of MPC. For a limited prediction horizon, NMPC often includes iterative solutions to optimal control issues similar to MPC. In linear MPC, on the other hand, these problems are convex; they are not normally convex in nonlinear MPC. This has problems for both the NMPC constancy theory and arithmetical solutions.

The numerical key of NMPC optimal control problems in one of the variables like direct single shooting, direct numerous shooting methods, or direct collocation, depends on the straight optimal control systems using Newton-type optimization plans. With the NMPC algorithms, it can be explored that the successive optimal control issues now close to each other. It is possible to save a significant amount of computation time with this advent and by correctly moving the guess from the previously determined optimal solution called the Newton-style Solution method. By using certain route following algorithms, the similarity of effective issues is further explored. These algorithms take little iteration to solve the most recent NMPC problem and never try to converge on any optimization problem.

NMPC is being increasingly introduced with slow sampling rates in some fields, such as the steps and chemical sector, and to fast sampling volumes in the automotive industry, with developments in controller hardware and computational algorithms. The NMPC is used in aerospace applications in real time to chart optimal terrain-following and avoidance trajectories.

8.3.5 Predictive Management Explicit Model

For certain systems, an explicit MPC (eMPC) enables the control law to be easily evaluated. EMPC is based on the problem of optimization, which can be called parametric programming, which is solved as a function of one or more parameters. The solution to the problem of MPC control optimization is recomputed offline, according to this theory. Since the control law is always in PWA form, the eMPC controller stores the piece-wise coefficients and affine performance for each subset of the state space (PWA). The coefficients and its certain parametric indications of all places and the PWA are constant in this sense. It is possible to obtain optimal control behavior applying the PWA coefficients saved for all regions, which is then reduced to evaluate the region first with both the current state and PWA evaluation. Since the total number of regions is limited and is uniquely suited for fast dynamics, and control systems, the execution of eMPC does not need significant computational possessions. In a turn, this significantly increases the memory requirements of the controller, and it is computationally costly to search the current control area.

8.4 Model Predictive Control Overview

The characteristics of the technology strategy for all controllers are shown in Fig. 8.1 [12].

1. The upcoming results for a given a horizon, termed the prediction horizon, are calculated by using the process model at each instant N. These projected outputs are based on previous inputs, previous outputs, and a future $u(k + N|k)$, $k = 0....N - 1$ control signals, which are the ones to be reached to the machine and measured for $k = 1, \ldots$ With $k + N$.
2. The set of predicted control signs is intended by enhancing a given criterion to sustain the steps as near as likely to the desired equilibrium point x_i. As expected an open-loop input, the past value of a closed-loop input u_i. The initial state $x0$ has been transformed into state xi of the closed loop.
3. The control horizon extends between k and $k + N_s$ and the prediction horizon extends between k and $k + N$ as shown in Fig. 8.1.

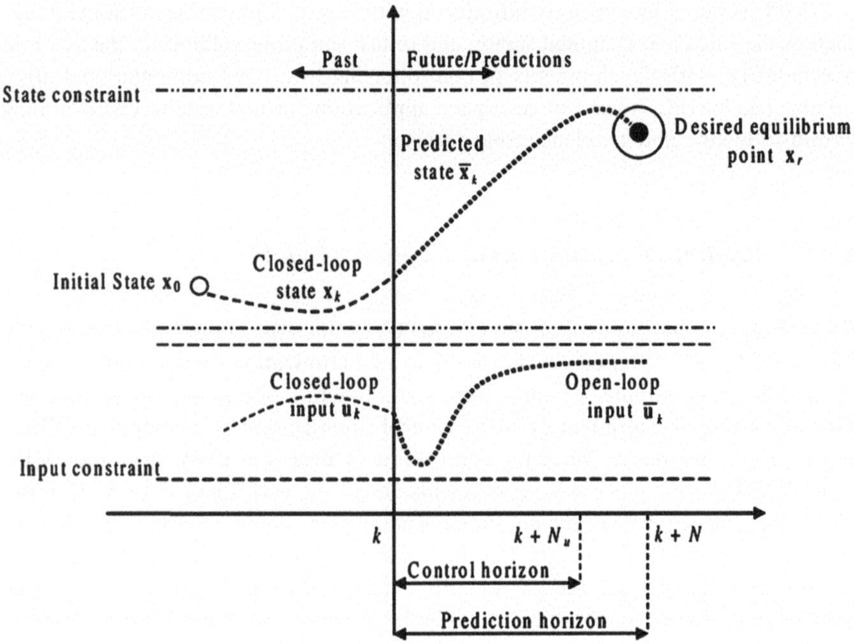

Fig. 8.1 Strategy MPC [11]

The basic structure shown in Fig. 8.2 is used to conduct this strategy. To forecast plant outputs based on past values and present values, a model is used. The optimizer measures these based on the cost function as well as constraints.

Within the controller, the elements model roles a significant character. To reliably predict future results and be easy to integrate and realize, the model selected should be able to imprison the steps dynamics. There are several modes of models used in various technologies since MPC is not a single methodology but rather a collection of distinct methodologies.

One of the most well-liked is the truncated impulse response model in the area of industry. This is very simple to accomplish since it involves just the output measurement when the method is excited with an impulse input. It can be used for multivariable processes, as it is very intuitive and widely accepted in industrial practices. The great number of parameters required is the main drawback of this model. The phase response model, where the input is a step, is very similar to this template.

Except for the multivariable case, the derivation of the controller is very straightforward, and the state-space model is, therefore, common in the educational study community. The key benefit of this model is a simpler representation of the parameters for stability and robustness. Another approach used in the academic research community is the transfer function model. In this model, as there are fewer parameters, it is more difficult for controllers. The key advantage of this model is the use of time gains and time constants that have vanished. This aspect makes this model more preferable in industries than the state-space model.

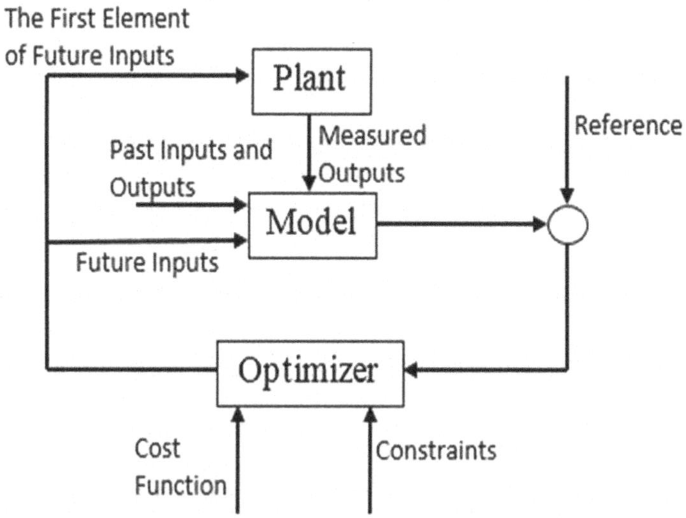

Fig. 8.2 MPC basic structure [13]

The optimizer, which offers control measures, is another fundamental part of the strategy. The smallest amount can be obtained as an explicit performance of past inputs and outputs, and the given potential suggestion trajectory if the cost performance is quadratic. A number of variables and the estimation horizon determine the size of optimization issues that do not require complex programming codes to be solved. In the sense of unconstrained cases, the timespan is greater compared to restrict and robust cases, and thus the bandwidth of the mechanism can be greatly decreased for the restricted MPC.

The control technique used in a driving vehicle can be regarded as an MPC strategy. The driver shall decide on the control action to be taken (accelerator, brakes, and steering) taking into account the characteristics of the vehicle, to follow the necessary reference route for a limited horizon of control. Only the initial management decisions take place at each point, and the process is frequent for the next control choice in a moving back horizon fashion. Past errors are used to make control decisions in the case of classical control systems such as PIDs [12].

8.5 Model Predictive Control in HVAC Buildings

New ideas and innovations have emerged in recent years to answer the crucial question of raising sustainable energy in buildings. Maximizing the application of on-site renewable energy sources while maintaining the thermal relief of the occupants inside the building by improving the energy efficiency of building envelopes. The study shows that commercial and residential generating account for around 40% of total chief energy in the United States and European countries and that the effect of ventilation and air-conditioning (HVAC) systems is largely preventable. A contrasting need for inside thermal comfort and construction energy demand seems to be present, but researchers and engineers are finding trade-offs between them.

Building occupants (i.e. Prosumers (Producers + Consumers)) play a vital role in improving energy efficiency, and their behavior is vital during building operations. In the future, to integrate these distributed resources and generation systems into smart distribution networks, the implementation of peak-shaving strategies and aggressive demand initiatives are expected to increase. Effective power management systems with higher control methods have been used to decrease the operating energy demand of buildings and disputes between their power demand and on-site age group, optimizing the manipulation of RES. More modern control methods ensure substantial energy savings in comparison to conventional control methods.

The key benefit of advanced control techniques is the accessibility of data, ICT advertisements, cost-reducing hardware, and energy management systems that pave the way for a significant volume of building-related data to be processed, stored, and analyzed. To this end, if data-driven systems properly process the information collected, it can provide vital knowledge of the actual performance of the building and the impact of respondent activities on the demand for construction energy. The study of these data can be a very useful opportunity, as a matter of interest, to turn the data collected into baked energy-saving strategies, and active demand techniques to improve energy efficiency in building and HVAC systems. As a result of this situation, a great deal of effort has been made to apply more advanced and prediction-based management strategies to boost energy efficiency in buildings. The competing aims are highly resolved by applying a model and prediction-based control approach that has the ability to search for optimal trade-offs.

Model Predictive Control (MPC) adds principles for feedback control and numerical optimization. By using the capabilities of energy storage and RES on-site generation optimization, it unlocks the possibilities. MPC can forecast future disruptions and provided requirements based on distinct control aims to predict the energy wants of the construction and improve its thermal behavior. Conditions are built-in explicitly in the optimization issue at each sampling stage.

The installation of wired, wireless sensors and integrated controls has increased rapidly in construction systems over the past decades. It is necessary to increase computational capacity to invent low-cost sensors and precise weather predictors to invent some achievable control strategies to boost successful building climate control. As several variables can be incorporated into the optimization of living space climate control when on-site production and energy saving are implemented at the home, it is very difficult to end up with a precise solution. The aims of a smart energy-saving and thermal comfort management system usually include the following:

- Minimizing the building's operating expense by achieving strong energy efficiency.
- Achieving a respectable degree of comfort by taking into account thermal air quality and visual comfort.

The American Standard for Heating Refrigeration and Air Conditioning Engineers Handbook offers a comprehensive overview of prevalent control methodologies for the construction of power systems. In the reason for its ease of a plan and low computational difficulty when formative the control signals, a method called classical control has been commonly used in building energy systems. In Rule-Based Controllers, the HVAC subsystems are regulated based on the inferential logic of if-then-else (RBC). For example, bang-bang controls are too common in old building systems without digital control, but Proportional-Integral-Derivative control is commonly applied in new modern buildings, where heating and cooling systems are fitted with electronically manage and variable frequency goes. There is typically no optimization at the level of the whole house, although there are sometimes highly sophisticated local controllers available. This means there will be a lack of an upper layer capable of maximizing each controller's set point. This is primarily due to the high complexity of an RBC controller which makes it technically difficult at the construction level to generalize the formulations.

Many researchers concentrated on the development and implementation of smart methods in the early 1990s for building control systems. Smart controls for the management of dissimilar subsystems of an intelligent structure could be optimally tuned using evolutionary algorithms. In this sense, the learning-based approaches of AI techniques are used to solve energy management problems in contrast to traditional methods. If it appears like the system is nonlinear, AI-based control provides a smoother operation for noisy or incomplete data. Once the system is educated, it can carry out predictions at a reasonably high speed. One of the most popular AI methods, ANNs, has been used at length for structure predictions and HVAC control techniques.

The issues of HVAC systems can also be resolved by Fuzzy Logic Controllers (FLCs). It offers potential options for coupling and incorporating the management of the HVAC device. Methods of optimization such as Genetic Algorithms (GAs) are used extensively to boost the parameters of other techniques of control. For the tuning of the classical controller and FLC parameters, researchers are working on the use of GAs. GA and the optimization of ANN models for the management of the HVAC system have defined the primary thermal parameters of a zone model based on capacity.

Classical control concepts are the focus of a growing number of surveys focused on analyzing the possibilities provided by the application of the technique. Thus, the principle mentioned in the chapter shows that MPC algorithms are an efficient way to increase energy efficiency in buildings. There are two major classes, one of which considers MPC to be only one of several potential control techniques, and the other is solely based on MPC. Demand Side Management (DSM) procedures are based on load forecasts and forecasts to clarify the energy management challenges. MPC is recognized as the most effective and gentlest method for a power management optimization system that denotes a house's higher degree of intelligence control. This is an example of the former one. The latter deals with the implementation of a thermal comforts building using only MPC or by integrating MPC with the artificial neural network (ANN).

8.6 MPC Models in Building and HVAC Systems

The creation and identification of an acceptable dynamic model are considered the hardest and time-consuming everyday job of the control design process, especially when using the MPC strategy. In the implementation of an MPC controller for buildings and their HVAC systems [14], there are two distinct important models that can be distinguished. They are the ones that control for used construction, and its HVAC system model is used for online optimization to test the thermodynamic behavior of the building used by the MPC. With the aid of the disturbance models, the behavior of the uncontrolled variables influencing the dynamic response of the system is predicted.

For the MPC controller implementation, the two models mentioned earlier are often needed with the replacement simulated construction model, and the high-exactness representation of the managed system required to close the control loop in the simulation is completed.

A system with all the different models toward the inside into MPC issues for construction and HVAC systems is shown in Fig. 8.3.

Figure 8.3a shows the MPC formulation with an alternative controlled system whose input is the output of a control-oriented model, that is, $u(k + 1)$. The output of the alternative controlled system model is $x(k)$ which is given as input to the control-oriented model. The gray boxes show the MPC formulations. There are many inputs to the control-oriented MPC model; they are cost function weights, problem constraints, output from disturbances model $v(k)$, and output from surrogate controlled system model $x(k)$.

On the other hand, in Fig. 8.3b, real controlled systems are replaced instead of surrogate controlled system models for both building and HVAC systems. Like in previous cases, gray boxes show the MPC formulations. There are many inputs to the control-oriented MPC model; they are cost function weights, problem constraints, output from disturbances model $v(k)$, and output from real controlled system model $x(k)$. The input to the actual controlled system is the output of the control-oriented MPC model $u(k + 1)$ and the output of the actual controlled system is $x(k)$.

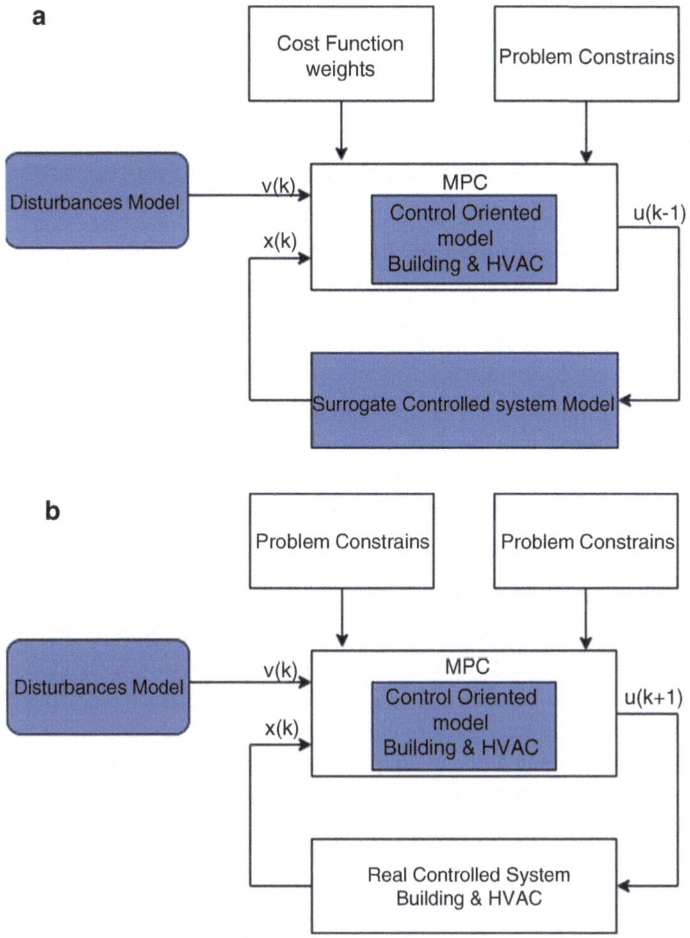

Fig. 8.3 (**a**) In the absence of a true regulated method, MPC formulation [14]. (**b**) MPC formulation for the actual implementation of a managed device [14]

8.6.1 Modeling of the HVAC System and the House

The modeling methods are divided into two separate types, according to the American Standards for Heating, Refrigerating and AC Engineers: the classical (forward) approach and the (inverse) data-driven approach. The classical approach is otherwise referred to as white-box models or engineering models. A thorough understanding of the different processes and connections of the system is presumed in the forward method. The key benefit of this forward approach is the framework does not need to be materially designed to test its performance. The classical method is highly appropriate for beginning forecasts of energy wants and the design of device loads in modeling a house's energy behavior. The most popular comprehensive energy instruments belonging to the forward approach methods are quasi-steady-state methods suggested by the norms, reduced-order models and updated bin methods [14].

ASHRAE, on the other hand, categorized data-driven models as seen in Fig. 8.4 into three broad categories, namely, calibrated simulation, gray-box modeling, and empirical approach. The real data that are monitored will be in the controlled plant.

- **Calibrated simulation models:** It is otherwise called white-box model. The calibration of data is done in the calibrated simulation. This is considered as the high-fidelity response model based on the substantial concept of measuring thermal dynamics and energy behavior of the entire building level or for sublevel elements. This is very similar to the forward process, except that the models are adjusted using the actual data monitored in the controlled factory in this case.
- **Models Gray-box:** It is the duty of the physical description of the system they serve to maintain this model and their parameters can be determined using system recognition methods. With the fitting parameters present in the gray-box models, it is possible to describe the dynamics of the physical system. The thermodynamics of a system is explained by an electrical network analog used in a Semi-Deterministic Physical Modelling (DSPM). The R electrical network analog is shown in Fig. 8.5. By turning to historical measurements, the model parameters are estimated. This is considered the reduced-order response model.
- **Black-box models:** The other name of black-box model is the empirical approach which is responsible for performing machine learning-based prediction strategy. Black-box models are built by adapting a model's parameters to the historical actions of the system without the need for complete knowledge of the system or the mechanism. To foresee process behavior where there is an ambiguous notion of a priori deterministic understanding of the physical relationship between input and output, as the black-box models are solely suitable for this.

The control-oriented construction model is nothing but the development of a virtual picture of the controlled system capable of accurately representing the

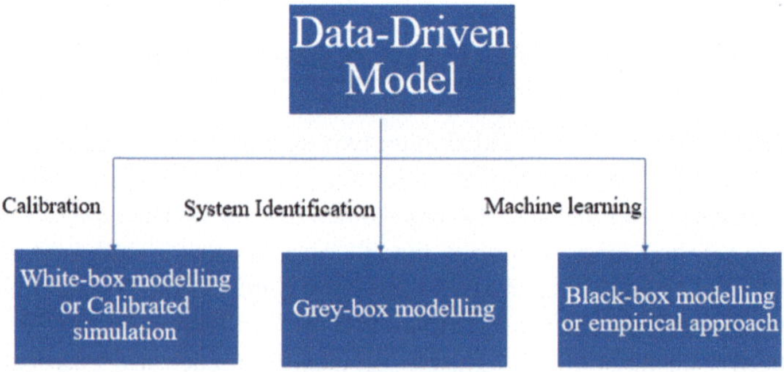

Fig. 8.4 Types of data-driven models

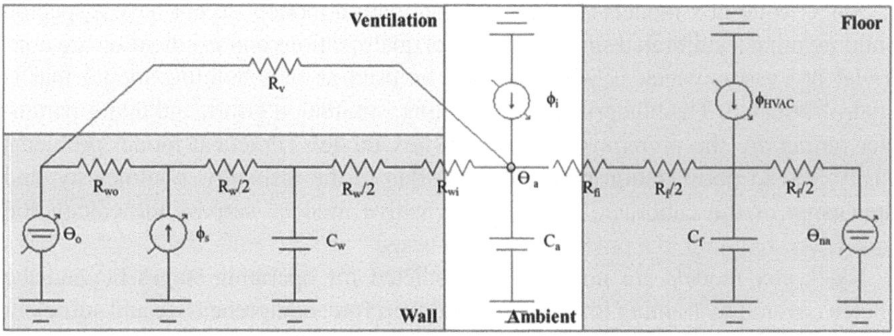

Fig. 8.5 R-C network example for MPC applications [14]

physics of these controlled processes so that the MPC strategy can be correctly implemented. This control-focused construction model must be sufficiently accurate to enable adequate forecasting capabilities and to achieve simple engineering steps that affect the powers of the managed building, but this must be very straightforward to ensure a fair measurement period for the optimization process. Since the model is a simplification of a specific physical system, it is important to reflect its response precisely enough to be successful in optimizing it. So, a data-driven approach supports most of the cases. In experimental case studies [14], historical data of states and disruptions which are detected by sensors can be used for this purpose.

The most important climate disturbances and occupancy-related disturbances, manageable keys such as the thermal energy supplied by the HVAC, are provided as inputs to the prediction and calibration models. Heating or cooling power offered by the HVAC, internal gains (occupancy or solar-related), solar radiation, and outdoor temperature is the usual inputs to the thermal model. Usually, the measured answer is the indoor temperature, but in some cases, it often includes the average temperature of the ramparts or other building mechanisms. Sensors are mounted on the walls, floors, ceilings, and other building components to determine the indoor air temperature. In certain situations, HVAC device states will be considered in the model. Apart from incorporating sensed temperature in the model, many other parameters can also be included to enhance the occupants' comfort. Examples of such parameters are light level and carbon dioxide level. By incorporating these parameters along with the temperature in the control-oriented model, user comfort can be enhanced.

Regarding thermodynamics, a construction can be used as a single zone or multizone. The number of internal nodes would correspond to the number of zones used to model the building's dynamics. A multizone building can be modeled from the viewpoint of the MPC and its use in the thermal management of indoor places, using a control-oriented building model. A distributed approach is taken in some other cases, where a separate zone is managed by various controllers. With this model having perfect prediction properties, the condition for the predictive controller's outstanding output can be derived.

Since white-box models or calibrated simulation models do not have a specific building model, calibrated simulation model qualifications and justification are nontrivial processes, which is why they can be used as in a building model that is control-oriented. The blueprint of the building, simulation effort, and major parameter tuning are the prerequisites of white-box models. Practical model problems easily lead to optimization problems that, due to the difficulty, nonlinearity, and dimension of the calibrated simulation-sensitive models, surpass the calculation time frame required in a practical controls usage.

Black-box models are not reliably predicted for operating stages beyond the choice covered by training information, and therefore, comprehensive and sufficient data preparation are required to ensure prediction correctness. These models can be designed easily and rapidly if sufficient data are available. To ensure accurate system representation, control-oriented models are often adopted.

Gray-box modeling is an efficient method of modeling a building's thermal response. In gray-box models for an MPC application, the key concern is to pick an acceptable model that is neither underfitted nor overfitted, and at the same time, embedding the information contained in the data must be slightly complex. To find the best model through an iterative process, a forward selection method for the most meaningful and sufficiently complex model is used.

By using the state-space sign of Linear Time-Invariant (LTI) systems, the gray-box model was formulated. The following equation shows the discrete state-space model.

$$x(k+1) = Ax(k) + B_u u(k) + B_v v(k) + Gw(k) \tag{8.1}$$

$$y(k) = Cx(k) + D_u u(k) + D_v v(k) + d(k) \tag{8.2}$$

where to,

$u(k)$ → vector of manipulated inputs or controlling variables

$v(k)$ → Measured disruptions vector influencing the device

$x(k)$ → The machine vector states

$y(k)$ → The output vector

$d(k)$ → Unmeasured random noise in measuring the states.

A, *Bu*, *Bv*, *C*, *Du*, *Dy*, and *G*, namely, the state matrix, the manipulated input matrix, the estimated disturbance matrix, the output matrix, the manipulated straight transmission input matrix, the established straight transmission disturbance matrix, and the matrix, the unmeasured random sound of the states. It is possible to measure these parameters contained in such matrices using system recognition techniques. Output is not a function of manipulated inputs, which for all construction applications results in a zero Du matrix similar to other manufacturing processes. The indoor air temperature can be determined for the outputs, $y(k)$, and the internal wall temperature is unmeasured. The formulation is often adopted in building applications in the formulation of state feedback, which is an alternative approach to the formulation of output feedback.

8.6.2 IoT-Based Architecture for an HVAC System

It is feasible to concentrate on the mathematical concept of the HVAC unit, and its energy elements. One such HVAC unit model is shown in Fig. 8.6.

PM2.5, carbon dioxide, temperature, humidity sensor, etc. can be integrated into the block diagram of the HVAC system to automate the lighting and fan system. In the microcontroller, the sensors are integrated. The microcontroller collects the sensed data and sends the data to the cloud by the Wi-Fi module. Smart electronic switch performs overload cutoff function. Power consumption measurements and ventilation system control. Air-conditioning remote control uses an IoT device to control the heating and cooling mode of the system. Mobile applications may be used to collect user preferred temperature values which are obtained as user feedback. The energy leadership policies are collected by the intelligent control manager. Peak electricity pricing, thermal comfort of the standard office space, room temperature, fixed temperature, difference between room temperature and target temperature at any time instantly, minimum temperature, maximum temperature, and the necessary variance in the comfort threshold can be included in the energy management policies. All the mentioned parameters are optimized by the intelligence control manager. The optimizer in the data analysis block will optimize all the data from the intelligent control manager, user feedback, smart electronic switch, air-conditioning remote control, and sensed data from microcontroller. To track the optimized data, the HVAC Unit is used. Storage and visualization of sensed data

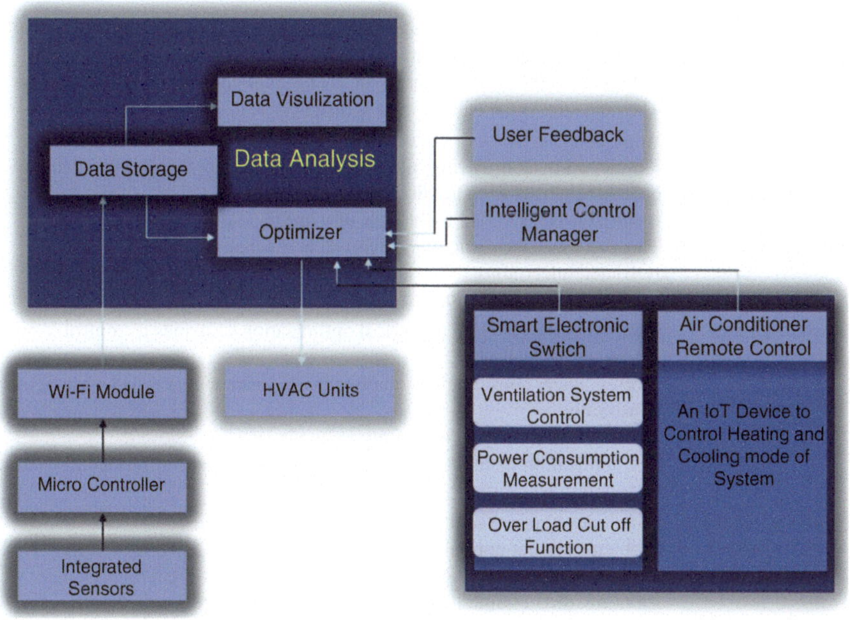

Fig. 8.6 Block HVAC machine diagram

will be taken place in the data storage and data visualization section of the data analysis block [15].

Energy management strategies are framed by believing that peak pricing is introduced at certain periods of the day [15]. The following is the policy on energy management.

The thermal soothe of the usual place of workroom can be intended on time using the following equation.

$$\mathrm{tml}_\tau = 1 - \left(\left|\mathrm{tmp}_\tau^{\mathrm{room}} - \mathrm{tmp}_\tau^{\mathrm{set}}\right|\right) / \mathrm{diff}_\tau \tag{8.3}$$

tml_τ—thermal comfort of the normal office space
$\mathrm{tmp}_\tau^{\mathrm{room}}$—room temperature
$\mathrm{tmp}_\tau^{\mathrm{set}}$—set temperature
diff_τ—difference between room temperature and desired temperature at time τ
Constraints in optimization problem,

$$\mathrm{tml}_\tau \geq \mathrm{tml}_\tau^{\mathrm{min}} \tag{8.4}$$

$$\mathrm{tmp}^{\mathrm{min}} \leq \mathrm{tmp}_\tau^{\mathrm{set}} \leq \mathrm{tmp}^{\mathrm{max}} \tag{8.5}$$

$$\mathrm{tml}_\tau \geq \mathrm{tml}_\tau^{\mathrm{min}} + \mathrm{ph} * \mathrm{tml}_\tau^{\mathrm{var}} \tag{8.6}$$

During regular hours, pH is set to 0 in Eq. (8.4) and 1 during peak hours.
Comparison table of typologies:

Classical model predictive control	Developed based on the concept of controlled invariance An initial step is the problem definition Followed by the absolute design of the optimal unconstrained control law The next step is the definition of settings for a dual-mode prediction
Robust model predictive control	Additive disturbances were considered in the initial process Developed by considering future realizations of uncertainty into account called the closed-loop prediction strategy It can also be extended in the case of a multiplicative ambiguity This discusses the difficulties involved in the control issue such as restricted uncertainty It is conservative in nature
Stochastic model predictive control	Belongs to the group of numerical optimization strategies to control stochastic systems It offers fewer conservative solutions that operate with a broader range of constraints Useful when the model ambiguity is stochastic The application includes chemical and process industries, production and distribution energy, risk management, finance, etc.
Nonlinear model predictive control	It includes iterative solutions to optimal control issues Problems are not normally convex The numerical key of NMPC in one of the variables like a direct single shooting, direct numerous shooting methods or a direct collocation depends on the straight optimal control systems using Newton-type optimization plans Used in an aerospace application

Predictive management explicit model	Based on the problem of an optimization called parametric programming It enables a control law to be easily evaluated It significantly increases the memory requirements of the controller, and it is computationally costly

8.7 Advantages and Disadvantages of Model Predictive Control

MPC increases energy savings and is very cost-efficient. It provides disturbance robustness and sifting in performing conditions. Within a boundary, it is capable of controlling multiple variables. It highly improves the steady-state response by decreasing the offset error. MPC predicts the upcoming disturbances and control actions. It enhances the transient response by reducing the rise time, setting time, and peak time. MPC also controls the slow-moving processes by using time postponements. It has the peak load shifting capability. These are the advantages of Model Predictive Control. The drawbacks are it is an accurate dynamic model and it is necessarily an optimization algorithm. It involves high computational cost. The problem becomes very difficult to resolve while uncertainties are considered.

8.8 5G and IoT Security Issues

Security is often defined as a method of maintaining a high level of confidentiality and integrity of the asset's information and making the information available whenever needed by the concerns of safeguarding a resource against physical damage, illegal access, or theft. On the other hand, it is also defined as a sequence of actions performed by means of any hardware or software tools following certain dishonest principal. Introducing constrained IoT devices and IoT technologies in such sensitive applications leads to new security challenges. IoT networks for its operations and execution are completely relying on the connectivity of myriads of devices. Hence, there is the highest possibility of being exposed to a security attack. The attack mainly ends up with alter, expose, destroy, disable, steal, or gain unauthorized information. The security services for communication over networks are provided by cryptographic security protocols. These services include data confidentiality, message integrity, authentication, and availability. In the context of 5G, DDoS attacks and signaling protocol-specific hacks of the past are considered as the most important 5G threats. Apart from these, there are many security issues that include lateral propagation, web application layer vulnerabilities, API security, and advanced persistent threats. For ensuring the privacy and security of users, it becomes highly essential to incorporate security issues in IoT-based architecture for all the applications. It is highly impossible to conclude a single security solution as IoT networks are heterogeneous in nature; so, depending upon our application, a security scheme has to be adopted to create a trust domain for users.

8.9 Conclusion

So far, the aim of this chapter has been to provide a lucid idea and a broad understanding of the use of MPC algorithms to normalize structure and HVAC systems. To accelerate the market penetration of MPC algorithms, it is important to clearly define the most capable structure classifications and users that can take advantage of their execution. It is important to implement these algorithms in a large building to experience the cost-effective technical solution of the MPC algorithm. The capital investment for the implementation of MPC technology is lower in large buildings compared to the reduction in operating costs. As with smaller houses, the advantages of the MPC application are more marginal. Passive and active storage techniques, elasticity of constraints, and alternation between occupied and unoccupied times are three parameters that are considered essential criteria for successful building control. A good candidate for the application of the MPC algorithm is considered modern corporate, institutional, and educational buildings, as it is very efficient for large buildings. Real-time implementation of IoT-based HVAC systems in buildings optimizes energy consumption, and, at the same time, it enhances user comfort. Thereby implementing MPC algorithms in HVAC systems will reduce the user complaints and energy bills. To implement demand response strategies directly, it is essential to incorporate the complete potential of MPC algorithms.

References

1. Basil Kouvaritakis, Mark Cannon, "Advanced Textbooks in Control and Signal Processing", https://doi.org/10.1007/978-3-319-24853-0, ISSN 1439-2232, Springer International Publishing, Switzerland, 2016, Model Predictive Control-Classical, Robust and Stochastic.
2. Zhao Zhou; Bart De Schutter; Shu Lin; Yugeng Xi, "Two-Level Hierarchical Model-Based Predictive Control for Large-Scale Urban Traffic Networks", IEEE Transactions on Control Systems Technology, Vol. 25, No. 2, pp. no: 2017, 1–13.
3. Xiaoguang Zhang; Yu Cheng; Zhihao Zhao; Kang Yan, "Optimized Model Predictive Control for PMSM Drives with Dead-Time Voltage Vector", IEEE Power Electronics Transactions, Vol. Number 36, no. 3, 2021.
4. Shuangxia Niu; Yixiao Luo; Weinong Fu; Xiaodong Zhang, "Robust Model Predictive Control with Improved Control Precision for a Three-Phase PMSM Motor", IEEE Industrial Electronics Transactions, Vol. Number 68, no. 1, 2021.
5. L. Yang; J. Lu; Y. Xu; D. Li; Y. Xi, "Constrained robust model predictive control embedded with a new data-driven technique", IET Control Theory & Applications, Vol. 14, No. 16, pp. no: 2020, 1–14..
6. Oldewurtel, F.; Parisio, A.; Morari, M.; Jones, C.N.; Gyalistras, D.; Stauch, V.; Gwerder, M.; Lehmann, B. "Energy-efficient building climate control using Stochastic Model Predictive Control and weather forecasts", In Proceedings of the American Control Conference (ACC), Baltimore, MD, USA, 30 June to 2 July 2010,
7. John Baillieul, Tariq Samad, "Systems and Control Encyclopedia" Reference Springer, https://doi.org/10.1007/978-1-4474-5058-9, May 2015.

8. Rel Guzman; Rafael Oliveira; Fabio Ramos, "Stochastic Model Predictive Control Heteroscedastic Bayesian Optimization", IEEE Robotics and Automation Letters, Vol. 6, No. 1, pp. 56–63, 2021.
9. Bernardini, D.; Bemporad, A. "Predictive control of stochastic constrained linear systems stabilising model", IEEE Trans. Computer. Autom. 2012, 57, 1468–1480 regulation.
10. Mady, A.E.; Provan, G.M.; Ryan, C.; Brown, K.N, "Stochastic Model Pre-dictive Controller for Integration of Building Use and Temperature Regulation", In Proceedings of the Twenty-Fifth AAAI Conference on Artificial Intelligence, San Francisco, CA, USA, 7–11 August 2011; pp. 1371–1376.
11. Mircea Lazar, "Model predictive regulation of hybrid systems: reliability and robustness", Eindhoven: Technische Universiteit Eindhoven, ISBN-13: 978-90-386-1823-4, 2006,
12. Eduardo F. Camacho and Carlos Bordons, "Model Predictive Control" Second Edition-Advanced Control and Signal Processing Textbooks, Springer Publication, ISBN:1852336943, 2007.
13. Ibrahim Celik, Mehmet Polat, "A Survey on MPC Widely Used Many Areas of the Industry", International Conference on Engineering and natural science, 2753–2759, 2016.
14. Gianluca Serale, Massimo Fiorentini, Alfonso Capozzoli, Daniele Bernardini and Alberto Bemporad, "Model Predictive Control (MPC) for Improving Energy Efficiency of Building and HVAC Systems: Formulation of Problems, Applications and Opportunities," International Energy Journal, MDPI, Vol. 11, no. 631, March 2018, https://doi.org/10.3390/en11030631.
15. S. Dhanalakshmi, M. Poongothai and Kaner Sharma, "Indoor Air Quality IoT Based and HVAC System Smart Energy Management", Third International Computing and Network Communications Conference (CoCoNET'19), Elsevier Procedia Computer Science, Vol. 171, 1800–1809 pp. no., 2020.

Chapter 9
Adaptive Multimode Decision Tree Classification Model Using Effective System Analysis in IDS for 5G and IoT Security Issues

P. T. Kalaivaani, Raja Krishnamoorthy, A. Srinivasula Reddy, and Anand Deva Durai Chelladurai

9.1 Introduction

Through the improvement in the 5G, IoT, increasingly more system gadgets interface with systems, which make network system information. Most assuredly, this information greatly supports network administrators recognize the network condition and consider its nature of administration (QoS). For instance, the system supports accepting the fleeting and appropriation, dissecting the presentation of system connections, hubs, and important conveyance and traded-off hubs. In this manner, arrange shortcoming areas, traffic perception, organize directing enhancement, assault recognition, and relief can be acknowledged as needs. System information assortment has become an urgent and fundamental part of organizing the board and assault discovery.

Although numerous large information handling procedures are helping huge network data information gathering process, the present network processing stands as yet standing up to difficulties. On the off-chance that every unique datum is gathered legitimately at each system hub, the efficiency of information handling and investigation is difficult to be accomplished. Compacting and reviewing information can discharge this issue; however, the exactness of further investigation may be

P. T. Kalaivaani (✉)
Department of ECE, Vivekanandha College of Engineering for Women, Tiruchengode, Namakkal, Tamil Nadu, India

R. Krishnamoorthy
Department of ECE, CMR Engineering College, Hyderabad, Telangana, India

A. S. Reddy
CMR Engineering College, Hyderabad, Telangana, India

A. D. D. Chelladurai
Department of Computer Science, King Khalid University, Gregar, Abha, Saudi Arabia
e-mail: anandevadurai@kku.edu.sa

S. Velliangiri et al. (eds.), *Secure Communication for 5G and IoT Networks*, EAI/Springer Innovations in Communication and Computing,
https://doi.org/10.1007/978-3-030-79766-9_9

influenced. In this manner, it is especially significant to examine efficient and viable information assortment strategies to gather as not many as conceivable system traffic information progressively and simultaneously to guarantee the nature of later information preparation and investigation.

One of the fundamental goals in designing systems of sensors for enormous scope information assortment is to accomplish longer lifetimes for the sensor organize organizations by keeping the vitality utilization at least while keeping up adequately high caliber and goal of the gathered information to empower significant investigation. Moreover, the setup of information assortment ought to be rearranged now and then to adjust to changes coming about from high framework elements (Fig. 9.1).

Sensor network and sensor systems experience a wide range of elements, including spatial and worldly change slants in the detected qualities that add to the ecological elements, changes in the client requests that add to the undertaking elements concerning what is being detected and what is viewed as intriguing changes, and

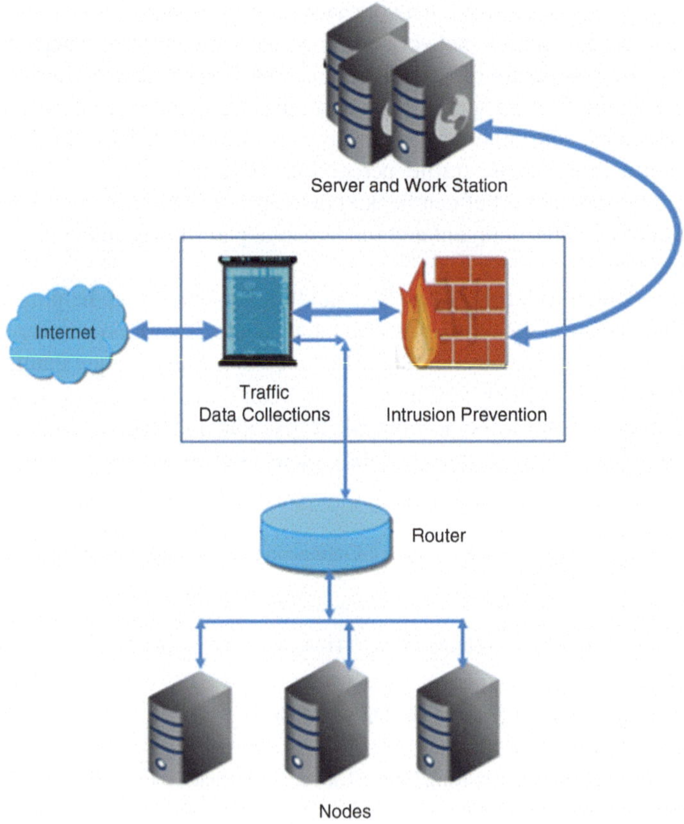

Fig. 9.1 Normal data collection in network

changes in the vitality levels of the sensor hubs, their area, or availability that add to the system elements.

One of the goals in arranging systems of sensor nodes for huge scope information assortment is to accomplish longer lifetimes for the sensor to organize arrangements by keeping the vitality utilization at least while keeping up adequately high caliber and goal of the gathered information to empower significant examination. Moreover, the design of the information assortment ought to be straightened out every once. The section discussed classifies the data collections on the networks, and tree models support the powerful predications.

9.2 Related Works

Several algorithms had been discussed for information collection on the network and argument number of the algorithms. The present system aims to manufacture logical strategies, the assessment of the nature of administration seen through the customers in the downlink of OFDMA remote cell systems serving to spill then flexible [1]. Originally depicted was the asset allotment issue in this work. The Quality of service used for FC may just through assessed in modernizations. This trouble is to be adjusted, recommending approximately sufficient feasibility conditions taking the multiclass processor sharing model, which licenses systematic assessment of the QoS.

Software-defined network (SDN) is another methodology of systems administration that accentuation on the system programmability and increasingly powerful control of the system [2]. Open Flow standard is the comprehensively utilized standard to execute the SDN organize. In any case, understanding the dynamic conduct of an Open Flow organize is yet to be tested since the data about the activity are appropriated over various system switches. The option gives an ongoing observing and representation of an Open Flow organize, including system geography and stream tables of each switch.

The authors proposed the flow broker and the individual layer leading to a different equalizer. It is then supervised and distributed over the supervised network controllers, and changes in the area supervised by each controller where the instruction is charged maintenance [3]. To deal with multiple space flow broker design load compliance, systems need to arrange execution improvement in programming.

Because of the different system prerequisites of utilizations, quality of service (QoS) mindful directing assumes a significant job in the systems. As of late proposed, asset distribution calculations center around the current traffic framework, which is not appropriate for dynamic systems. It is misused the estimation of stream framework that enables the plan to adequately lessen the all-out parcel misfortune and at the same time raise the system throughput [4]. Along with these, scientifically planned QoS-mindful asset reallocation in software-defined network (SDN) systems dependent on the traffic forecast has also been considered.

This innovation assists with tackling a noteworthy number of notable designing issues powerfully and richly as they give programming characterized concentrated system control [5]. An SDN controller can be stretched out with an application that viably fills for solid needs and gives adaptable administration of system streams. This opens an extraordinary number of chances for a great deal of system security issues, for example, keeping up of benefits in a legitimate way, parting control and information planes, and assaults identification and alleviation. In this work, the chances of SDN for a "survival" moderation during DDoS assaults the heap adjusting issue is considered.

Security Intelligence Factor discount is an information security plan for the time being. While the development of security at the edge of contacts is bleeding and expanding their use of the safety specialized training to be credible, there is no nomenclature of safety norms among ordinary customers, making them the most delicate association in conjunction with any. In this way, advanced criminals have advanced hacking methods that can be used to make money and information from the general population [6] to contribute and take great measures to prolong the investigation of boredom. These numerous routes were considered in Dubai and Sharjah to integrate a comprehensive of security audits, most of which are safe for either the vulnerable or the usable infant to touch the weakest.

The wireless sensor network is an important part of the population that can be used in many fields to monitor environments and meetings after the environment [7]. Important, self-integrating dynamics of topology and tight resources, however, have unique characteristics that make WSNs vulnerable to attacks. Various systems are programmed to effectively detect the types of WSNs attacks. Now the first protocol of this study will provide an overview of WSNs and categorize WSNs attacks based on layer upon layer.

The enormous popularity of cell phones in the general public has roused the improvement in versatile learning applications. Notwithstanding the advantages concerning instructing and preparing, the current learning applications, despite everything, need to address issues and provoke identified with the turn of events, reuse, and structural normalization [8]. It investigates Software Product Lines (SPL), and it has been completed to utilize the approach through the advancement of portable learning applications. An SPL centers on programming reuse, and it has been effectively applied for specific areas.

Information collection is a significant stage in the web of things (IoT). The productive method to gather information from the IoT condition is one of the difficulties for the eventual fate of IoT. In this chapter, an information assortment framework for IoT is concentrated as a first step. This framework is based on Bluetooth Low Energy (BLE) innovation to advance the gathered information from the information gatherer to the information passages and proposes the utilization of PDAs as information authorities. We propose the first model based on this framework [9].

Hadoop information expectation [10]. Using the open-source MapReduce system derived from smaller terminals for transport projects of various sizes depending on the subsystem. MapReduce can record the entire design of execution and processing to improve the system. Interim, Supported Vector Recession Meaning

(SVR) Expected Traffic Stream. Its exciting assumptions used to execute traffic can match growing expectations. This design and mass traffic investigation situation seems to indicate the stage where programs can handle problems and achieve high traffic congestion forecast accuracy.

The general idea and execution of training focused on informal community information stockpiling and examination frameworks are talked about in this chapter. The requirement for such an idea originates from the way that right now little consideration is paid to the internal structure of such frameworks [11]. The proposed framework composes the way toward gathering information, filling the database from outside sources, and investigating the gathered information. The primary issue of the social event divided information from various sources is additionally tended.

Mobile ad hoc network (MANET) is turning out to be one kind of major next-generating remote system. It effectively experiences different assaults because of its particular qualities. To assess and gauge the security of MANET continuously and cause this system to respond appropriately, one pledge option is to integrate second-line location tools with job seekers for security to identify attacks in MANETs. That is the most offensive sign in this system. Collecting data related to reliable security [12] in MANETs has been distributed for over 20 years, and then proposed for different requirements of audit authorization tools.

The Long-Term Evolution (LTE)/LTE-progressed (LTE An) organize offers the kind of help that has paid billions of customers with its superior transmission, improved range effectiveness, and reduced inactivity beyond traditional settings. Be that as it may because of all its IP-based diversified design, it all experiences new security risks despite the experience [13]. In this way, LTE/LTE became a basic requirement for a system to play out a quick and accurate security prediction. Assault detection LTE Bar/LTE—is a system security appraisal where security important information is reported to be an essential for sequencing and information investigation.

Intensive research has led to large economic losses from cyberattacks and network security interventions. Network security is typically reflected on the website with some relevant data that can be stored on the computer. By observing the data, this website detects system interference and measures its security level, called learning and security-related information. Observingly, the first step in detecting network interference is to gather security-related information [14]. Obtaining information related to gift design and implementation of compatibility security on network systems based on multiple networks.

The internet is becoming a very difficult and sophisticated attack on the network. Major research arises to diagnose security risks and measure the security of the internet. Data collection is important for all types of methods for detecting cyberattacks and calculating its security, and the literature does not contain the complete study involved in the collection and research of security-related data on the Internet [15]. We have proposed to evaluate the current diagnostic methods for service flood and worm attacks denial assessment using the requirements for their effectiveness.

The final determined critique outlines a list of open issues and further research identification.

Adaptive network data collection, data detection, and data integration and essence systems, governance, and machine learning application-layer tunnel detection [16]. Economic data storage introduced the current research results of my research team based on this type of network. The reputation of the website is well-publicized. In the process of data collection and analysis of security-related network data, we collect and process certain information, such as environmental feedback system positive data, in an attempt to achieve security detection results that are becoming more accurate.

These challenges include issues that need to be addressed to increase resource poverty and data security and privacy, increasing the vulnerability of systems in need of mobile. As a defensive mechanism to prevent accidental security threats, mobile devices can attack vital resources on mobile devices [17]. Concepts of the need for autonomous computing and computer security should be combined with mobile device security measures to create a framework for the adaptive security framework. Automation and support for a built-in Android emulator began to be developed to determine the security hardware and software sources of the critical device users.

Domain generation protocol (DGA). Three-letter system and tunnel detection. The program detection method consists of two main parts: the rule-based domain filtering based on the machine extraction model of the proposed extraction framework [18]. Rule-based DGA domain name filtering can reduce some explicit mining to reduce the amount of data processed by machine-based learning detection thereby improving detection capability. The common feature is the integration of previous research and complement, multilayer and multilevel support, and multi-feature extraction results into the extraction framework.

Artificial intelligence (AI) has always vowed to get natural features for network traffic and to do some work on traffic forecasting. In this study, the AI-based lightweight adaptive measurement system (ALAMM) was used by SDN to reduce transportation cost measurement and improve measurement accuracy [19]. First, the transport network uses the model in an apologetic way to predict traffic. Based on the traffic forecast results, an adaptive method is used to determine the traffic sample frequency. Second, we send a switch calibration source to measure the outflow and approximate flow measurement connection.

Accurate and effective attack detection is a great traffic efficient active. In this thesis, the project is meaningful in finding techniques for multiplication attacks and reversible drawing [20]. This action allows the detection system to handle large amounts of network traffic congestion. Recovery Sketch based on the collected traffic network: We will use the rest of the theorem Chinese, and then constantly monitor the abrupt change to map the one-to-one request packet and the multiplier attack flow identifier as the response set. The nondetection mechanism is also powerful and useful because it does not make full use of the basic characteristics of complex traffic operations recording and multiplication attacks [21].

Various network backgrounds need smart awareness for data collection methods. To summarize, further analysis of the data term that can be published in this issue may be inaccurate. Present methods of data collection are not flexible and smart. Still, some direct issues occur in the network data collection processes. Mice flow flood module within a spice fiction network settlements that lead to a huge success. On the one hand, multi-pocket model algorithms are used in spy fiction c network situations.

9.3 Materials and Methods

In this section, the multimode decision tree classification introduces for data collection a design established based on the abovementioned research method. The present algorithm-based data collection by needful is its working procedure. Then, we present algorithm support to improve our network data collections, monitoring, traffic, and network management and intrusion detection, our prototype system that can be passed based on our platform. The data collector collects security data from any network node and sends it to the data server. Here, many problems need to be solved. For example, our equipment contains a security-based data collection process that allows network node devices to play with different types of networks to access multiple network systems so that they can replace each other based on user requests. Complex network demographics and many network systems have the potential to support strong transformations across different networks.

The architecture of adaptive multimode decision tree classification model-based data collections. Also, Fig. 9.2 depicts the stages of the proposed approach in detail.

9.3.1 Multimode Decision Tree Classification

Multimode Decision Tree Classification sequentially builds weak learners. Originally, it was designed in such a way that at every step, the sample distribution was adapted to put more weight on missed data collection and less weight on correctly collected samples. The final prediction is a weighted average of all the weak and stronger support for data collections.

Classification segments are designed as independent variables that take the whole number of random values, with the prediction error measured based on the weak data collection classification in the network. Recession trees depend on variables that are ordered to take continuous or discrete values, and typically measured by strong data collection are erroneous with prediction values observed and predicted on the network.

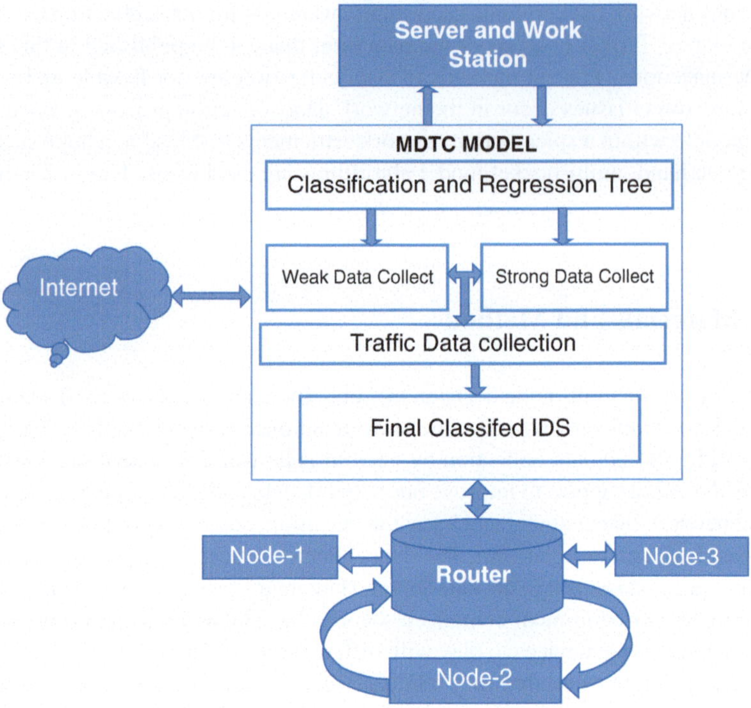

Fig. 9.2 Architecture of adaptive multimode decision tree classification model

9.3.2 Algorithm: Multimode Decision Tree Classification Model (MDTC)

Step 1:

Set two weak data weight and strong data counters
Ct1 = 0 and Ct2 = 0

Step 2:

For all new example, i.e., for individually network connection

(a) Set weight for weak and strong data (Cmt) = 1.
    Randomly example it consuming Poisson distribution.
(b) Repeat the weak classifier as,

If it suitably classifies the section,
Ct1 = Ct1 + cmt; // Ct1, Ct2 is counter varibles
Estimated weighted data collection level (CL) is updated by,
CL = ct2/ct1 + ct2;
Weight of connection (Cmt) is simplifiedby,

Cmt = cmt (1/2(1 − CL))
If it failure classifies,
Ct2 = ct2 + cmt;
stimated weighted data collection level (CL) is updated by,
CL = ct2/ct1 + ct2;
Weight of connection (cmt) is simplified by,
Cmt = cmt (1/2*CL).

Step 3:

Finishing weight (α) for weak data collection classifier is,
α = log (1 − CL/CL).
This weight returns the suggestion of example for detecting intrusion.

Step 4:

The process to strong data collections
Else if max (Cnt)

Ct2 = ct2 + cmt;
Estimated weighted data collection level (CL) is simplified by,
CL = ct2/ct1 + ct2;

Weight of connection (cmt) is simplified by,
Cmt = cmt(min (1/2*CL)).

Step 5:

Concluding weight (α) strong data collections
α = log (1 − CL/CL).

This weight intrusion reflects the involvement of a diagnostic model or feature.

This model is based on two tree models, regression tree and classification tree algorithm-based processing. Both are supporting weak and strong data classification. An algorithm is a time-dependent system called a program. Factors in the present tense can be generated based on factors from earlier times. Factors can be sample pocket interval, sample time interval, and so on (Fig. 9.3).

Now, in this section, we compare the exhibition review of the proposed project. All issues with service and data collection quality and protocol redirection were discussed as well as supply review and network improvement.

9.3.3 Data Collection

The data collection Center well defines support for data collection from various networks. Sometimes interface with various users, 4G/5G network, Ethernet, and mobile users.

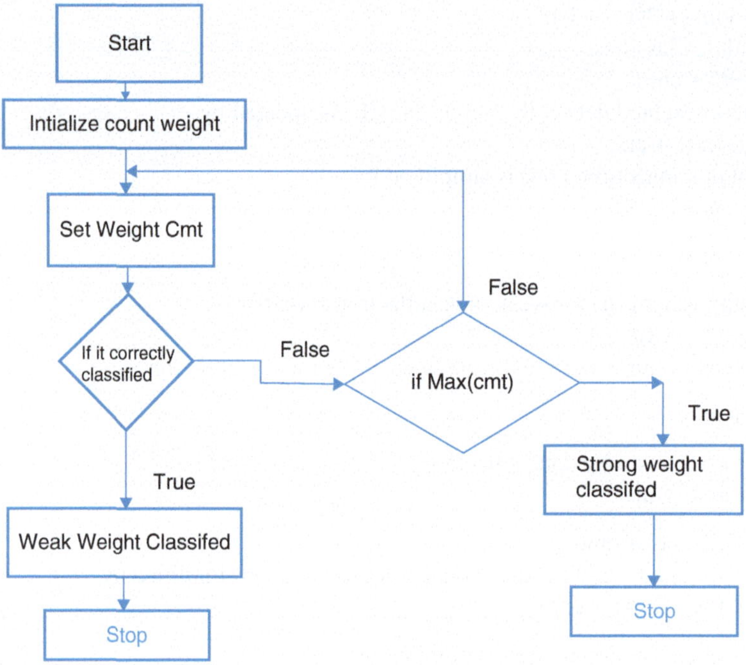

Fig. 9.3 Shows the flowchart of MDTC model classification

All user data and data collection center interface. A set of data that the network administrator would need. Support for secure information gathering across the entire network. Security oriented in the following steps. Contains event-related and clear security event data to gather the critical phase of detecting network attacks and interference. However, how to gather this information to standardize security-related data and how the network system's measurement results use these data are still an open issue for the integrated network.

9.3.4 Data Monitoring

The MDTC model gives the packet care component flows controller around the message a great mentality. Receives time in the presence of flow from the parcel and switch scalable message in the web system message. In the controller plane, connection exposure module, topographic refresh module, and device handlings are present. Topographic changes control system detection control connection exposure module, topography executive module, and device control module retrieve system topography, hardware, and connections data from pockets in messages.

9.3.5 Traffic and Network Management

The module collects information from selected nodes. For each TRAF sound quality, different information can be adapted to calculate the amount of additional information to reduce in the test stream. The nature of TRAF Network C is different. They can be pocket rate, streaming transfer, and so on. Allow network topography data and network evaluation results.

9.3.6 Intrusion Detection System

For companies to achieve the greatest perception of service, this consistency and service quality is a key factor to be determined. Arrived, however, entered the firewalls and detection methods data center. This (MDTC) model is currently used for infiltration detection methods and malicious behavior support.

9.4 Result and Discussion

Multiple and various problems reflected by the matrix are found in each section of the chapter. For authentication problems, the best and most efficient solution is proposed and the protocol is implemented while simulating using the popular network C#.Net. Both design and development were done using this language. We discuss several issues related to the MDTC model supported for the data collection on sensor networks.

9.4.1 Testing Systems

The plan is to run a laptop Windows10 operating system, using the Intel Core i3 CPU @ 3.30 GHz and 4G RAM at the end of the year. Four types of traffic cases are considered in the MDTC model as mentioned below. Our active virtual network accepts the fat branch, which is built by the ten simulation tools of the .NET Framework.

9.4.2 Traffic Bags

Explain the process of Detection System or DDoS Attack that DDoS attack is mainly requested to contain through the attack, Thoughtful increase counter-DDoS attack, and link flood attacks. Direct DDoS attacks and data packets are similar to re-Elective amplification negative DDoS attacks, and Different data packets come from the same stream. Therefore, we can detect a large number of direct DDoS attacks and reflection amplification negative DDoS attack identification amplifiers (Fig. 9.4). By way of a result, classify the present attack due to the different types of DDoS attack distribution established due to the traffic flows. Now, in our tests, we reflected on the following four traffic cases.

1. No traffic detection on the network.
2. Various flows lead to the conquest that is within a detailed network settlement.
3. Those successes are traces of a large number of traffic circulations.
4. A large number of some attacking nodes are said to cross certain network connections by being completed.

9.4.3 *CPU and Memory Performance*

Consumption in terms of the system and the four scenarios (Figs. 9.5 and 9.6). The link shows the usage and memory intake of the system by any flow controller. About 4% of our systems use controllers more systematically than in daylight. Although our recycling does not take up more than 60 MB of memory than in daylight, this network is suitable for data collection systems. Therefore, this system can store resources and memory resources. Figures 9.7 and 9.8 show the change in system usage and memory consumption as a sketch base and MDTC model system. The business system has a multisystem utilization and memory-swallowing rate system by at least 49% to reduce resource consumption.

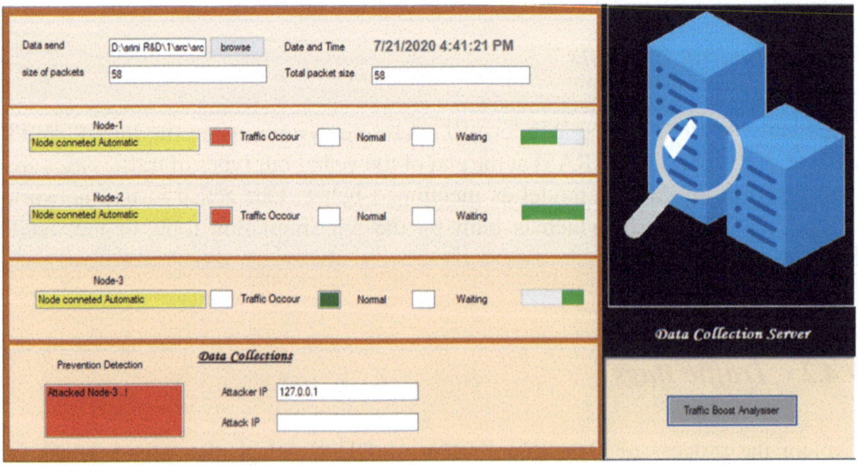

Fig. 9.4 Attack detection information collection server

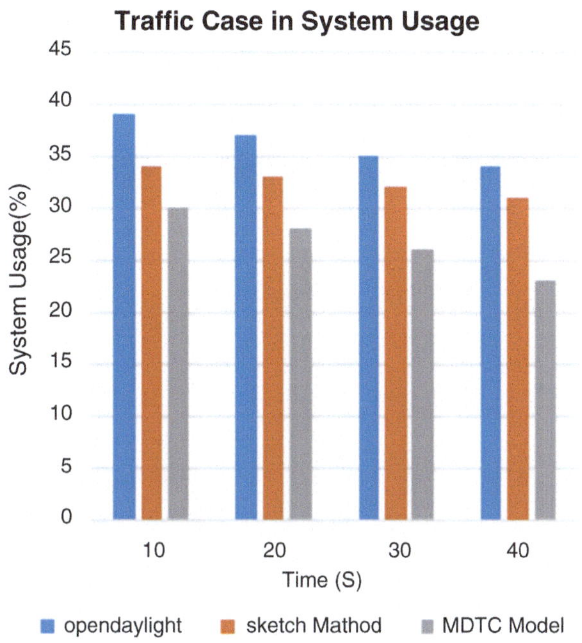

Fig. 9.5 Traffic case in system usage

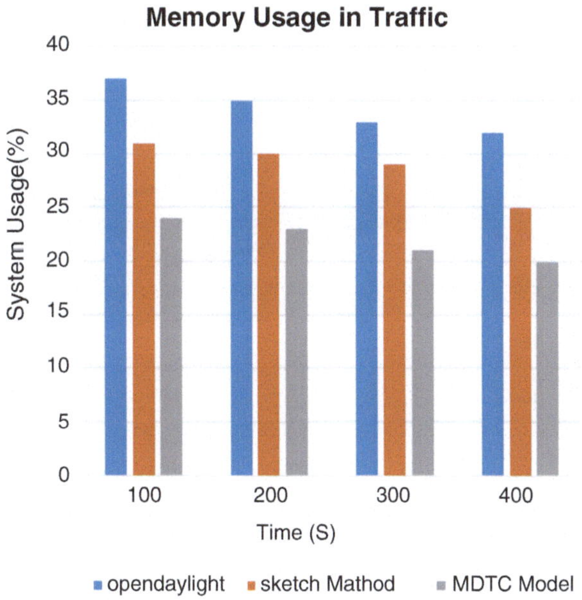

Fig. 9.6 Memory usage in traffic

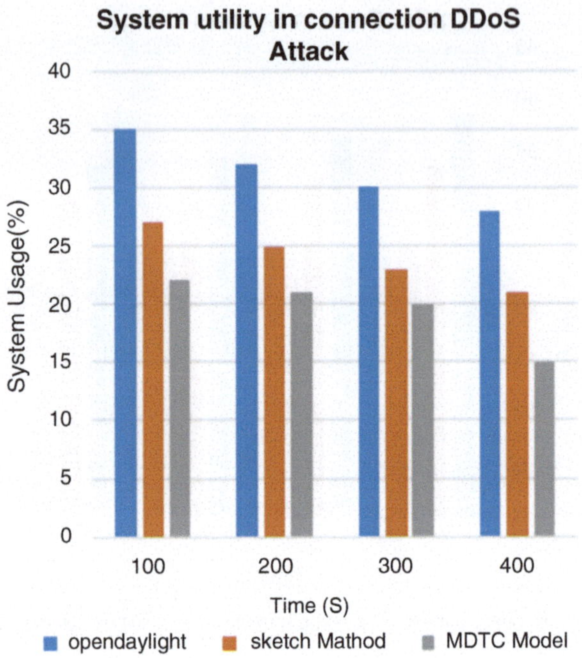

Fig. 9.7 System utility in connection DDoS attack

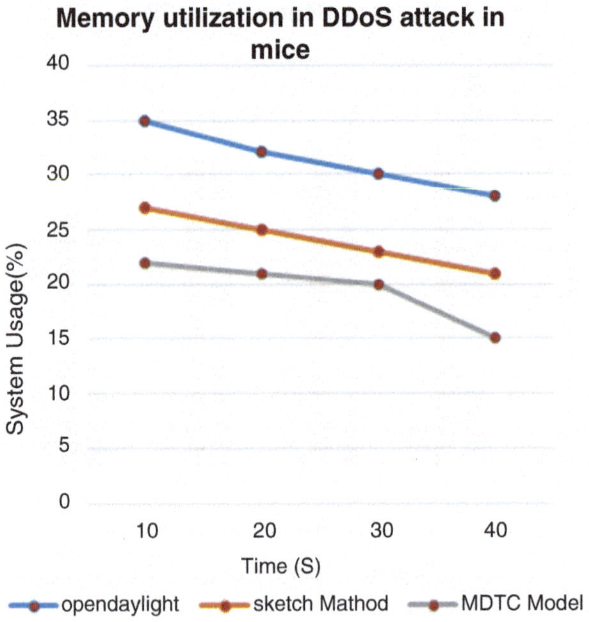

Fig. 9.8 Memory utilization in DDoS attack in mice

9.4.4 Storage Performance

We correspondingly confirmed the resource consumption of our system MDTC model basic collection types. If the flow elephant is distributed according to the ratio of the flow to the experimental movements. Tested in several cases: (1) 90% elephant movements and 30% mice movement; (2) 60% elephant movements and 50% mice movements; and (3) 20% elephant movements and 85% mice movements. As shown in Figs. 9.5, 9.6, 9.7, 9.8, and 9.9, the flow model in the system therefore greatly decreases storage consumption.

9.4.5 Threat Perception

Classification of Our system threat such as feedback movement, mouse movement and link motion-based DDoS attack have been discussed and shown in Fig. 9.8 and diagrams. It may be based on our company's threat feedback performance model collection bag data, the number of game bags, and the total two predictions of movement are determined in these three attacks. Now, Elephant Game DDoS Attack can bypass attack detection in the Elephant Game suite system. A miscalculation of the MDTC model can quickly make the elephant understand and realize the DDoS attack against multiple domain tables. Flow-based DDoS attack after mice ensues

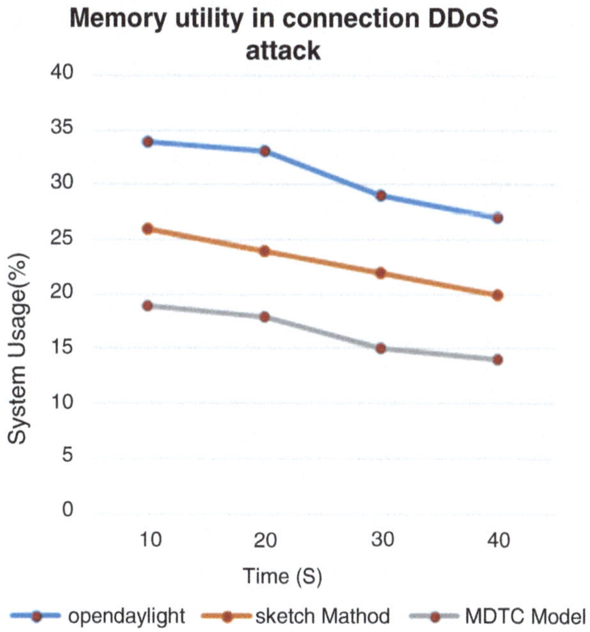

Fig. 9.9 Memory utility in connection DDoS attack

with a huge number of packets that need to enter switches or routers belonging to the new flows, and these flows usually have no adjacent packets sounding like flows extensions. Hence, the amount of new movement calculated in the prototype system be able to simply be detected (Tables 9.1 and 9.2).

A single image of the sketch-based system can detect movement, and then mouse movement cannot detect it. Therefore, it may be unfamiliar with this type of DDoS attack. Game-based DDoS attacks may be the basis for connection flood attacks, such as blocking certain nodes in the network. The MDTC system usually uses gateways or host nodes to detect possible attacks. DDoS for the normal delay is measured based on various controls such as attack time spent; we provide data collection on the network. All of these methods of magnification are used to calculate the average DDoS attack. The whole time it wanted a pocket from source to the scope of scope.

9.5 Conclusion

We recommended a sensitive data collection system and information system as the basis for the MDTC sample classification. It uses a system estimation and valuation model to evaluate organize node and connection states for choosing legitimate nodes for information collection. Exam centers adaptively study flow qualities, so economics, efficiency, and information data collection guarantee low asset usage. We evaluated our structure and directed the progress of analyzes based on a testimonial-idea model rather than its exhibition and different techniques as far as CPU/memory utilization, stockpiling use, flow size recuperation, and danger recognition. The test results show that system memory usage, storage consumption, and detection of IDS had the best performance in MDTC model classification. If the proposed system generates an accuracy of 6.76%, the false-positive rate is 96.41%.

Table 9.1 Memory consumption performance

	Opendaylight		Sketch base method	MDTC model
Network condition	Element movements	No. of movements	Packet movements element	The element of packets
Elephant	4969	1139	4972	4998
Mice	579	5260	590	620
Link flood	4950	1090	323	252

Table 9.2 Comparison of collected data & Capability of threat perception

Network condition	Opendaylight	Sketch-based method	MDTC model
Elephant	Yes	Yes	Yes
Mice	Yes	No	Yes
Link flood	Yes	No	No

References

1. "Analytical evaluation of QoS in the downlink of OFDMA wireless cellular networks serving streaming and elastic traffic," M.K. Karray, 2010, IEEE Transactions on Wireless Communications, vol. 9, no. 5, pp. 1799–1807.
2. J.A. Wickboldt & P.H. Isolani, "Interactive monitoring, visualization, and configuration of OpenFlow-based SDN," 2015 in IEEE International Symposium on Integrated Network Management.
3. D. Marconett & S.J.B. Yoo "A software-defined network controller architecture for multi-domain brokering and reputation," 2015 Network and Systems Management, vol. 23, no. 2, pp. 328–359.
4. Dai & W. Ying & J.M. Wang, "SDN-based multi-class QoSguaranteed inter-data center traffic management," X. 2014 in Proc. of IEEE International Conference on Cloud Networking.
5. M. Belyaev & S. Gaivoronski, "Towards load balancing in SDNnetworks during DDoS-attacks," 2015 in Proc. of Science & Technology Conference.
6. Velliangiri, S., Sekar, R., & Anbhazhagan, P. (2020). Using MLPA for smart mushroom farm monitoring system based on IoT. International Journal of Networking and Virtual Organisations, 22(4), 334-346.
7. T. M. Khoshgoftaar, R. Zuech & M. M. Najafabadi, C. Wheelus, "A sessionbasedapproach-foraggregatingnetworktrafficdata—The SANTA dataset," 2014 in Proc. 14th Int. Conf. Bioinf. Bioeng., Boca Raton, FL, USA, pp. 369–378.
8. R. D. S. Torres & G. M. Waku, E. R. Bollis & C. M. F. Rubira, "A robust software product line architecture for data collection in Android platform," 2015 in Proc. 9th Brazilian Symp. Softw. Compon., Archit. Reuse., Belo Horizonte, Brazil, pp. 31–39.
9. S. Moussaoui & O. Nouali, A. E. Boualouache & A. Derder, "A BLEbaseddatacollectionsystemforIoT," 2015 in Proc. 1st Int. Conf. New Technol. Inf. Commun., Mila, Algeria, pp. 1–5.
10. Z. Xing & W. Jiewu, H. Chunjing & F. Wentao, "Usertrafficcollectionand prediction in cellular networks: Architecture, platform and case study," 2014 in Proc.4th Int. Conf. Netw. Infrastruct. Digit. Content, Beijing, China, pp. 414–419.
11. X. Hou & S. Zhu, J. Liu, H. Wen & G. Sun "Trafficcollectionandanalysis system," 2016 in Social Computing, W. Che et al. Eds. Singapore: Springer, pp. 229–234.
12. Z. Yan & G. Liu, A. W. Pedryczc, "Data collection for attack detection and security measurement in mobile ad hoc networks: A survey," 2018, J. Netw. Comput. Appl., vol. 105, pp. 105–122.
13. Velliangiri S., Kasaraneni K.K. (2020) Machine Learning and Deep Learning in Cyber Security for IoT. In: Kumar A., Paprzycki M., Gunjan V. (eds) ICDSMLA 2019. Lecture Notes in Electrical Engineering, vol 601. Springer, Singapore. https://doi.org/10.1007/978-981-15-1420-3_107
14. Z. Yan & Y.L. Fu, H.Q. Lin, "Adaptive security-related data collection with context awareness," Journal of Network and Computer Applications, 2019 vol. 126, pp. 88–103.
15. Z. Yan & X.Y. Jing, W. Pedrycz, "Security data collection and data analytics in the Internet: A survey," 2019, IEEE Communications Surveys and Tutorials, vol. 21, pp. 586–618.
16. Karthikeyyan, P., & Velliangiri, S. (2019, July). Review of blockchain based iot application and its security issues. In 2019 2nd International Conference on Intelligent Computing, Instrumentation and Control Technologies (ICICICT) (Vol. 1, pp. 6–11). IEEE.
17. AakashAhmad, "Adaptive Security for Self-Protection of Mobile Computing Devices," Computer Science, 2019.
18. Zheng Yan & Huaqing Lin, Gao Liu, "Detection of Application-Layer Tunnels with Rules and Machine Learning," Computer Science, 2019.
19. Sheng Qi, Liuwei Huo & Lei Miao, Houbing Song & Dingde Jiang, "An AI-Based Adaptive Cognitive Modeling and Measurement Method of Network Traffic for EIS," Mobile Networks and Applications, 2019.

20. Zheng Yan, Xuyang Jing & Witold Pedrycz, Qinghua Zheng & Jingjing Zhao, "A reversible sketch-based method for detecting and mitigating amplification attacks" Computer Science, 2019.
21. Velliangiri, S. and Karthikeyan, P., 2019. Hybrid optimization scheme for intrusion detection using considerable feature selection. Neural Computing and Applications, pp. 1–15.

Chapter 10
Enhanced Role-Based Handover Control Algorithm for Efficient Multimedia Data Communication Performance in Vehicular Network Using IoT Security Issues

P. T. Kalaivaani, Raja Krishnamoorthy, A. Srinivasula Reddy, and S. Velliangiri

10.1 Introduction

With the modern annual growth rate, VANETS offers ideal protection and rich tourist comfort. However, the protection issue may appear to be an ad hoc network or a specific VANETS requesting situation involving standards. The latest developments in this research improve the security of the architecture and system technology, therefore. It is also equipped with suggestions for its protection issues, summarizing the main results. Security VANETS says human life is constantly threatened, and the most important security concerns in traditional networks include confidentiality, availability, and maximum security, with integrity. An attacker cannot delete vital statistics in a way that neither has changed. However, VANETS security proposes the ability to determine the use of pressure obligations while still maintaining driving privacy. Information about the vehicle and driver in the message change is important safe and is overdue to be replaced in excess. The timing can be motivated to have disastrous consequences, such as vehicle collisions (Fig. 10.1).

Connecting a car allows you to connect to the cloud and add your information to sensor readings, website visitor records, and so on. For unique car downloads, this

P. T. Kalaivaani
Department of ECE, Vivekanandha College of Engineering for Women, Tiruchengode, Namakkal, Tamil Nadu, India

R. Krishnamoorthy
Department of ECE, CMR Engineering College, Hyderabad, Telangana, India

A. S. Reddy
CMR Engineering College, Hyderabad, Telangana, India

S. Velliangiri (✉)
Department of Computer Science and Engineering, B V Raju Institute of Technology, Medak, Telangana

S. Velliangiri et al. (eds.), *Secure Communication for 5G and IoT Networks*, EAI/Springer Innovations in Communication and Computing,
https://doi.org/10.1007/978-3-030-79766-9_10

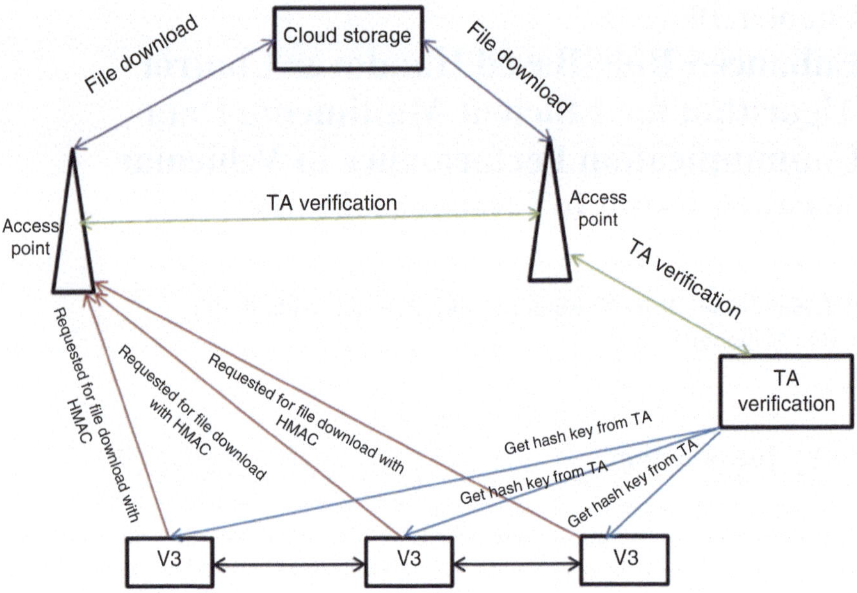

Fig. 10.1 VANET-based cloud storage system [29]

may be followed. This allows anonymous sharing of records between customers and keeps them secure. As the cloud has announced that website traffic, channel conditions, accidents, and weather conditions are according to vehicle sensor statistics, the application server in the data mixes the fact that it comes from the vehicle can assemble facts. It primarily improves information reliability by filtering biased or false records based on the client's data amount.

The deployment of the complete safety tool VANETS is very difficult in exercise. Protection against VANETS violations is often essential to shock. Besides, the individual in-vehicle networks are generally very dynamic, arriving and deviating from the motor and short connection intervals. The use of Wi-Fi media, in addition to its dynamics and excessive liquidity, allows for maximum open, broadcast-enabled Wi-Fi communication in VANETs vulnerable to attack. VANETS password attacks have been categorized later. In addition to the well-known network protection issues, certain security requirements drive a rise in VANET's exact characteristics, including excessive mobility, dynamic topologies, and general disconnects during fast connections. In addition to controlling certificates, these unique features perform security issues, such as forming, detecting, and protecting the feature's actual institution. Art responses' recent works can be sourced in the next section, primarily based on the nature of protection.

10.2 Related Work

However, VANET is difficult for video data transfer because it is a hostile environment for any type of vehicle network data transfer, especially video transmission (massive data packet loss, node movement, etc.). Therefore, researchers are also focusing on improving and integrating video communication. Unfortunately, video transmission [1] performance evaluation is not an easy task. In-vehicle networks are an urgent challenge in providing high-quality video streaming road safety and infotainment services. To provide a high-quality video stream, forward error correction (FEC) is one of the most popular ways to ensure video quality by creating fake packet streams. However, some factors, such as limited wireless network resources, can cause FEC problems with streaming video data [2], which is highly energetic.

Packets for copied data may also be more than network eligible. Therefore, in this study, the transmission mechanism is used to protect the transmission [3]. The main problem with the redirect technique is the delay. A multidirectional solution [4] is the controversial value of interference based on a generalized algorithm to minimize transmission rates and acceptable delay results. A fraudulent trading partner can waste valuable bandwidth [5] if frequent communication interruptions occur in the transmission of deceptive video clips. However, power outages challenging mobility characteristics during communication video services in VANET have been deployed in packet-loss vehicle networks [6].

However, there are several issues [7], such as data packet delay or packet loss, most of which need to be addressed. Due to the VANET-specific environment, the website was developed by traditional protocols such as the real-time transfer protocol (RTP), which is difficult to apply. Date [8] research work offers several new protocols to solve this problem. In non-emergency times, various multimedia applications can be used and streaming mobile cars to improve downloaded videos' quality. During the streaming of the video stream, vehicle video [9] may be switched from one RSU to another, affecting cognitive quality.

5G mobile specification is expected to meet the additional requirements for extra delay and reliability and accommodate mobile information traffic proliferation, supporting various high-speed applications and services. SDN implementations are gaining momentum, and, administratively, they are still full of [10] scalability and performance issues due to detailed networks.

Downstream completion time: Instead, switch buffers are needed for subsequent enhanced power applications to provide more reliable packet delivery than what is provided by electronic packet switch (EPS) networks. They were expected to be a 5G network. Higher throughput sub-degrees reduce the chance of losing the number of packets running on the opposite end using each packet's end-to-end deadline and the EPS network [11].

The excellent performance facilitates asymptotic behavior to prove the development of hundreds of networks [12] in collaboration. The optimization solution is a value factor, a centralized or distributed deployment, that embodies each information center and is compromised between good network load value and information center resource cost. Best multipurpose model of economic experts to achieve the balance between network and information center [13].

This relationship is linear to standard mobile phone networks, so small virtual cells significantly increase network capacity. Also, it seems that the best virtual small cell density for users is the price of direct macrocell connections and small-cell macrocell relay links, anytime and anywhere [14]. A riot was also a term, and we looked at specific combinations of streams and data streams. For a given topology, its elimination of variability can provide programming techniques using specific LAN speeds combined with a planned scheduling comb [15].

Pre-licensed transportation (5th generation (5G)) cloud radio access network transmission (C-RAN). The design of next-generation electronic networks requires high capacity and minimal overhead long-distance networks. The analytical community specializes in Ethernet-based packet-switch networks [16], as investment objectives may apply to mathematical multiplication gains, infrastructure restructuring, and ultimately price reductions.

Rapid mobile data systems require a rising framework of technical prominence and goodness combined with improvements in the field and the quality of multilevel transportation services. Therefore, in that case, VANETS has actively witnessed [17] much attention to the latest developments in telecommunications. In recent years, many intelligent transportation systems (ITS) can reduce traffic accidents and congestion and provide passengers with primary and useful information. However, this is why it is there once the design of these standards has significant machine sorting communication response parsing [18].

Main course management mechanism of dynamic cluster-based physical examination action, to ensure dynamic clustering work under the changing, in the long run, take account of vehicle quality and implementation rules of the clustering algorithm, sporadic. The road conditions are planned. But the method of determining this era is still controversial and can add to the computational burden of cluster heads. With SDN, the management logic can be a hierarchical network structure, as it is far from the underlying infrastructure to control the controllers in layers [19].

The coordinator is implemented with a higher level of software-defined network (SDN) management design than the world's domain controllers. The controller and resource essence between them also play an important role in the background system [20]. Attempts to support computer information and domestic loose 5G wireless hotspots to make networks own beneficial resources are available. It often seems limited to cooperative caches and various constraints on limited storage capacity, content, placement, and content quality distribution [21].

However, human activity trends and cluster habits were not considered in the ancient user's quality model [22]. As a keyword, user suspension, user arrival, and the opportunity to leave a contribution are derived in the 5G small cell network [23], which is very hot to evaluate user performance quality. Fifth-generation wireless communication systems have become an important framework to meet various network application pieces' needs in different application scenarios. Spread infrastructure network multiple logic private networks are divided, and the wireless network supports various services. However, this is an end-to-end tablet that is very rapidly deployed [24] at the heart of a multidomain wireless network infrastructure.

However, in many cases, it is tracked that the attacks are manufactured by positioning controllers in distributed denial of service networks or (DDoS) attack methods in VANETs, protected from vehicles, different types, such as blackhole, replay, and masquerade [25, 26]. Although there is no information or control plane overloaded or delayed at the management layer, SDN and mixed supports in 5G configurations are topics implementing security services in VANET networks. This system effectively meets the VANET requirements for low latency and good response [27–30].

This section describes video broadcasting based on various methods defined: broadcast storm avoidance mechanism (BSAM), vehicle position prediction algorithm (VPPA), weight-constrained shortest path problem (WCSPP) are defined. But these methods are not well known for video broadcasting communication in VANET. So, the enhanced role-based handover control algorithm (ERBHC) will be used to implement the broadcasting communication.

10.3 Materials and Methods

In recommending ERBHC to protect and share video content, a specific group is based on people's cloud-based automotive system and their role in password schemes. It must implement a role-based access control mechanism for accessing videos stored on cloud platforms. A vehicle equipped with a computer device becomes a mobile terminal, and the network becomes a vehicle ad hoc network (VANET). Streaming video content is sent to the internet in an abbreviated form for viewing by real-time viewers and vehicles (Fig. 10.2).

This vehicle is equipped with ample internal storage. Therefore, the VANETS node is powerful and transfers continuous video data to other vehicles and roadside receivers. VANETS uses a hybrid framework that includes both infrastructure and temporary structures. It is a temporary vehicle-to-vehicle (V2V) communication and vehicle-to-infrastructure (V2I) access point (roadside device). These roadside units (base stations) are then connected to the internet and provide the vehicle's

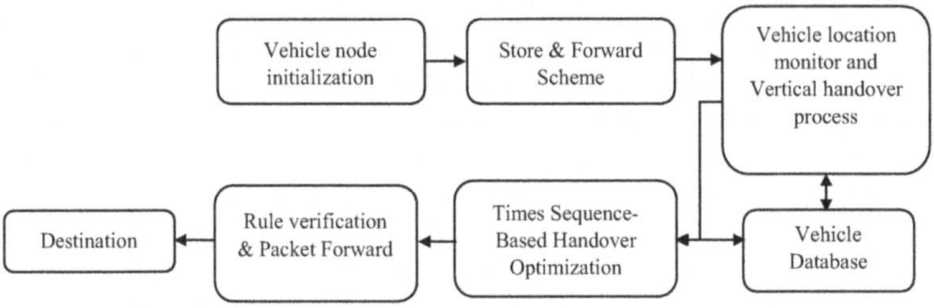

Fig. 10.2 Proposed system block diagram

necessary services. The success of VANETS depends on the availability of a sufficient number of vehicles with roadside infrastructure and wireless communication equipment.

10.3.1 Store-and-Forward (SF) Scheme

Two network measurements, large packet delay and packet loss, affect the video quality of the recipient. In streaming video, video frames are broken into many small data packets and sent to the network. If a disproportionate percentage of lost packets is combined with error correction, the receiver will not recover the frame. The decoder will also discard it at the receiver if the packet's arrival time exceeds the corresponding frame's reproduction period. Therefore, delay and packet loss may cause video distortion. The vehicle's high dynamics can explain the challenges of VANETS video streaming. High speeds and limited vehicle safety often lead to connection breakdowns and even network partitions. This is the amount of delay, and it takes for the car to catch up with other cars in advance and reconnect to the network and day of day. The technology employs buffering store-and-forward (SF) data packets and sends them at the next opportunity in the catch-up phase. Streaming applications generate many data packets, so it may also be a relay vehicle that can cause severe packet loss and link buffer overflow.

Encryption for Video Broadcasting

Input: Parameters from the message (m, d, m, n), Public Key P,
Raw Text m Output: Encrypted textual content (content_1, content_2) begin
Represent the message m as a point P in $N(R_p)$.
Select okay $\in P[1,n-1]$.
Calculate $\text{content}_1 = m_p$.
Calculate $\text{content}_2 = m_p + n$
Return (content_1, content_2) end.

Decryption for the Video Broadcasting

Input: Parameters from the encrypted message (m, d, m, n), Private key d, Encrypted textual content (content_1, content_2)
Output: Raw Text m begin
Calculate $m = \text{content}_2 - d_{m^2}$ dC1 and extract m from M.
Return (m) end.
End

To protect eavesdropping (video), the node should let both parties share the mystery or exchange the contract's public key. As a result, the conversion of encryption keys for ad hoc networks, which translates very quickly, can be resolved due to whether it is attractive, given the secret of pre-negotiation. Among the proposed technologies, one of the heterogeneous implementations is based mainly on the encryption method used to protect the video recording reporting service when tourists are filming.

Public and private keys: This technique uses two keys. The public key is publicly available, and in very simple messaging, the sender uses the recipient's public key to encrypt the message. Privacy is used to encrypt the recovered message.

10.3.2 Vertical Handover Process Based on ERBHC

The vertical handover process is based on some procedure steps. These procedure steps are given subsequently:

Step1: Discovery of Network

This part should accumulate the software layer, fact hyperlinks, and social delivery layer together by the facts from different layers. These layers provide RSS, bandwidth, ultrafast, throughput, user potential, and networks. This record is processed and used to make handovers during high-quality selections of the ERBHC algorithm if the network is deployed.

Step2: Handover Triggering and Decisions

The handover decision is to decide this mechanism if you want to keep cellular equipment and modern radio to access the community or switch to all other networks for everyone. Switch selection depends on different metrics collected during the switch discovery phase. Handover can be described primarily based on the transmitted symbols from the target neighbor BS over the serving BS through a given threshold from modern times. This part makes relevant choices when switching. In this area, the choice of the target radio access network is based on several criteria.

Handover Execution

Its basic function is to get the right to enter the broadcast through the association of the target and the mobile wireless terminal and realize the vehicle is switching optimization system. This degree of switching also includes authorization and ongoing transfer of user statistics during active sessions. This is the last segment of a switching system where modern consulting is being transferred to a brand-new access network to run the area.

Input: The number of mobile nodes
Output: Effective handover
Begin()
Deploy vehicle ad-hoc network with a limited number of RSUs and vehicles
Root node selection
for ($i = 0;i==N;i++$)
if (Range($N(i)$) > Range($N(i + 1)$))
$N(i)$ = root NodeEnd
Source select Destination Node(i)
If(Root node==Node(i))
Source select root node for communication
Else
Source sends a request to another root node.

When handover occurs, users can choose heterogeneous networks with their preferences and network performance parameters. User choice can be beneficial to the network, consumer application necessities (real time and non-real time), service type (voice, data statistics, and video), and quality of service. It will be a fixed technology network traffic. An effective way to increase the user's preference for wireless environments (such as reporters) can also optimize the vehicle in next-generation technology wireless networks can consider switching.

Times Sequence-Based Handover Optimization

Time series runs with lower network partitions along with higher consistency requests and use different mechanisms for calculations.

Input: compute a node, Neighbor Partition NP
Output: Routine RE
Step 1: Start
Step 2: Select a node
Step 3: network partition neighbor PN
Step 4: Start PN Timer NP
Step 5: While H-Timer is running
Receive Reply Partitioning RP.
Remove Node details and location details.
Update Neighbor table NT for every entry
NP (i) = PN {NODE – Id, P, Q} ||
End
Step 6: Stop
Where,
P-source,
Q-destination
PN–Partition Node

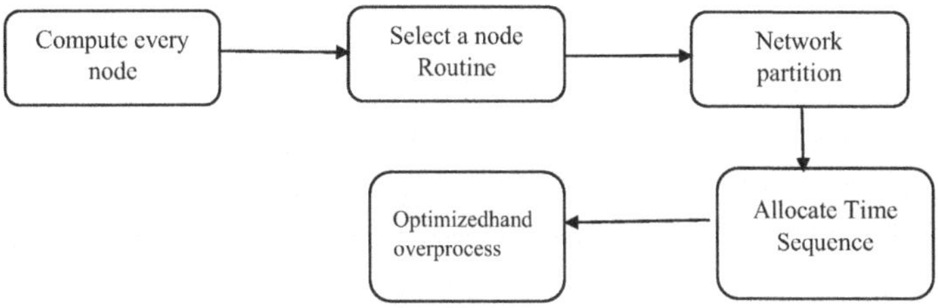

Fig. 10.3 Times sequence-based handover optimization

ID-Identify node

The device calculates the time series as the time required for the item between the time changes in the influence rise and its last notification (Fig. 10.3).

10.4 Results and Discussion

The basic network is formed in the sensor implementation of the VANET framework. The network can be formed to monitor traffic without any congestion or problems. These nodes are formed and considered vehicle nodes, base stations, and others. In portability proof, the development of one center to another region is authorized from an area that can use the Tool Command Language (TCL) content slogan target.

Table 10.1 describes the resources needed in this proposed system. This section compares the existing broadcast storm avoidance mechanism (BSAM), vehicle position prediction algorithm (VPPA), weight-constrained shortest path problem (WCSPP), and the enhanced role-based handover control algorithm (ERBHC).

Figure 10.4 shows a comparison with the proposed and existing method. The existing methods are BSAM in 87.4%, VPPA in 88.6%, WCSPP in 91.2%, and the proposed method ERBHC in 94.6%.

Figure 10.5 describes the graph that shows the throughput performance between the proposed and the existing methods. Based on the existing algorithms (BSAM in 178 in kbps, VPPA in 235 with kbps, WCSPP in 290 with kbps, ERBHC in 327 with kbps), a small number of packets are transferred over a more extended period, and more packets in the ERBHC are transferred over a longer period.

The transmission ratio is used to estimate the quality of the network. It defines the ratio between the target receiving packets and the source-generated packets. This can be achieved using the awk script that creates the trace file and the result.

$$\text{TR} = \text{Received packets} / \text{Generated packets} * 100$$

Figure 10.6 describes the transmission ratio in the percentage between the proposed and the existing systems, comparing prevention methods in the transmission ratio. When there are 60 nodes, the PDR of ERBHC is compared to BSAM in 75%,

Table 10.1 Simulation parameters

Parameters	Value
Data rate	2 Mbps
Application type	Constant bit rate (CBR)
CBR interval	1.0 (s)
Simulation time	350 s
Simulator	NS-2
Data speed	20 m/s
Total energy	500 J
Number of vehicles	4

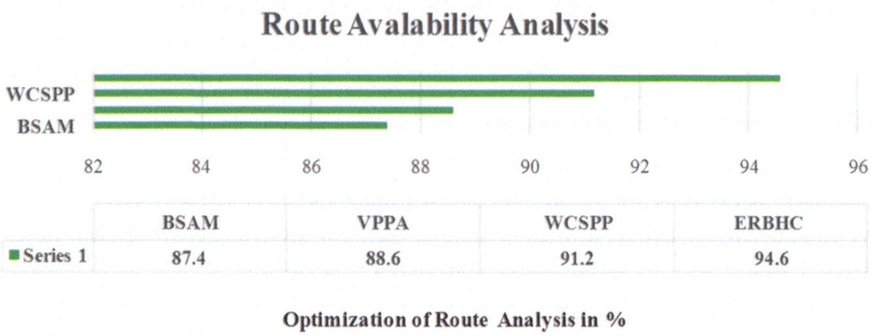

Fig. 10.4 Optimization of route analysis

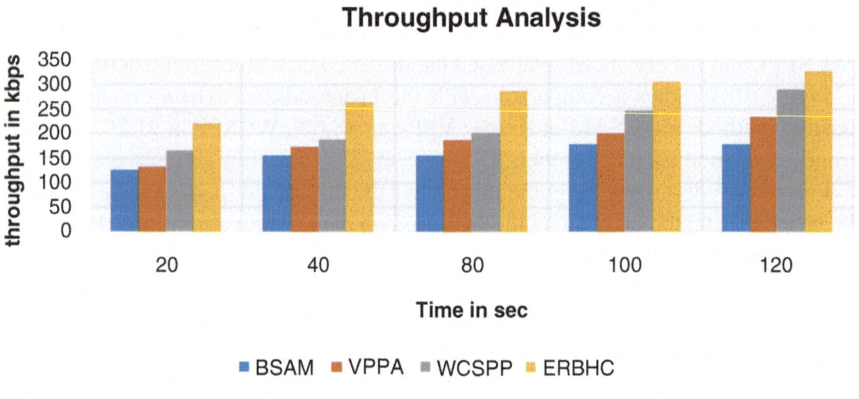

Fig. 10.5 Throughput performance

VPPA in 85%, WCSPP in 87%, and ERBHC in 92% with a transmission ratio value of 95%.

Energy consumption in Fig. 10.7 illustrates the clearly understood planned method of consumption. When the number of nodes is high, the transmitting energy is high. In this research work, the total energy consumption is to reduce the total

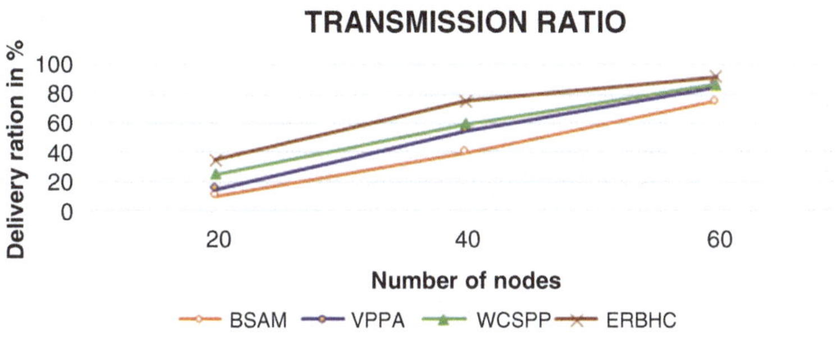

Fig. 10.6 Transmission ratio

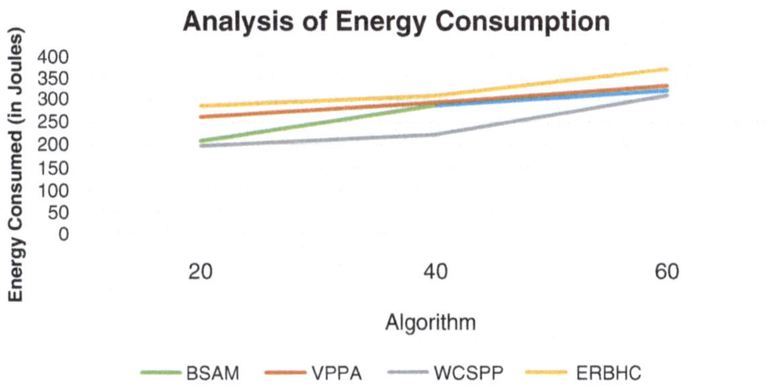

Fig. 10.7 Energy consumption

energy. Energy consumption can be used to calculate the system's efficiency to reduce energy when sensing the data.

It is proposed that the vertical handover is based on the ERBHC method to provide a reduction in the number of data packet transmission delay times. It is clear from Fig. 10.8 that the proposed vertical handover can be calculated for 50 nodes and compared to the proposed system. The handover delay shows the delay time for BSAM with 342 s, VPPA with 385 s, WCSPP with 287 s, and ERBHC with 217 s.

10.5 Conclusion

ERBHC dynamic transmission range and rate are message types at all reasonable times. This strategy consists of two parts, which include congestion detection and congestion control components. The congestion detector measures the usage level of the channel and compares it with a predetermined threshold. If the channel usage level exceeds the threshold, then congestion occurs in the network. In the congestion control device, the transmission distance and speed are adjusted. Multi-target

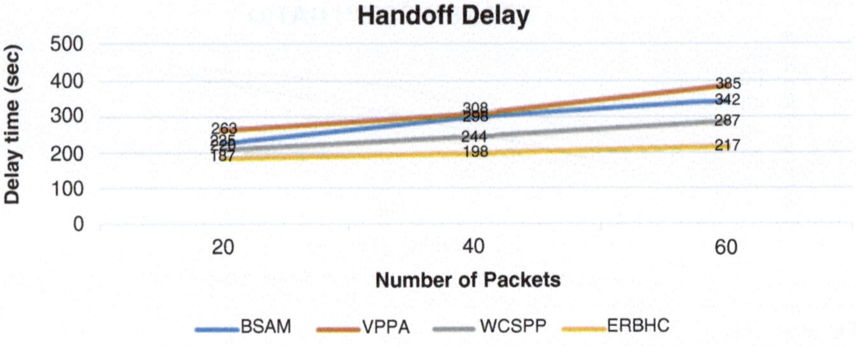

Fig. 10.8 Analysis of handover delay

traffic monitoring and secure broadcast are used for secure video transmission without any congestion and implement an improved switching mechanism. The objective function of this research activity is the minimum delay. It is an NP problem to get a reasonable value for speed due to transmission range and high computational overhead. ERBHC's research area of privacy in video-reporting services can provide evidence of cryptographic verification in digital video reporting services in vehicle-to-vehicle communications. Deploy VANETS security's state-of-the-art facility security infrastructure to secure video and reporting systems. The proposed ERBHC analyzes the optimization of route analysis value is 94.6%, throughput performance is 327 with kbps, transmission ratio is 95%, energy consumption is 372 J, and handover delay is 217 s.

References

1. Marzouk, H., Badri, A., Sahel, A., Belbachir, A. K., & Baghdad, A. (2019). New Simulation Model for Video Transmissions in VANETs using EVALVID framework. 2019 7th Mediterranean Congress of Telecommunications (CMT). https://doi.org/10.1109/CMT.2019.8931395.
2. Vineeth, N., &Guruprasad, H. S. (2015). Performance analysis of network coded video streams in VANETs based on mobility models. 2015 IEEE International Advance Computing Conference (IACC). https://doi.org/10.1109/iadcc.2015.7154692.
3. Smida, E. B., Fantar, S. G., & Youssef, H. (2018). Video streaming forwarding in a smart city's VANET. 2018 IEEE 11th Conference on Service-Oriented Computing and Applications (SOCA). https://doi.org/10.1109/soca.2018.8645770.
4. Smida, E. B., Fantar, S. G., & Youssef, H. (2017). Video streaming challenges over vehicular ad-hoc networks in smart cities. 2017 International Conference on Smart, Monitored, and Controlled Cities (SM2C). https://doi.org/10.1109/sm2c.2017.8071838.
5. Xie, H., Boukerche, A., &Loureiro, A. A. F. (2015). A Multipath Video Streaming Solution for Vehicular Networks with Link Disjoint and Node-disjoint. IEEE Transactions on Parallel and Distributed Systems, 26(12), 3223–3235. https://doi.org/10.1109/tpds.2014.2371027.

6. Velliangiri S., Kasaraneni K.K. (2020) Machine Learning and Deep Learning in Cyber Security for IoT. In: Kumar A., Paprzycki M., Gunjan V. (eds) ICDSMLA 2019. Lecture Notes in Electrical Engineering, vol 601. Springer, Singapore. https://doi.org/10.1007/978-981-15-1420-3_107
7. Joshi, J., Jain, K., Agarwal, Y., Deka, M. J., & Tuteja, P. (2015). COMIP: Cluster-based overlay and fast handover mobile IP system for video streaming in VANET's. 2015 IEEE 3rd International Conference on Smart Instrumentation, Measurement and Applications (ICSIMA). https://doi.org/10.1109/icsima.2015.7559039.
8. Mammeri, A., Boukerche, A., & Fang, Z. (2016). Video Streaming Over Vehicular Ad Hoc Networks Using Erasure Coding. IEEE Systems Journal, 10(2), 785–796. https://doi.org/10.1109/jsyst.2015.2455813.
9. Smida, E. B., Fantar, S. G., & Youssef, H. (2017). Predictive Handover Mechanism for Video Streaming in a Cloud-Based Urban Vanet. 2017 IEEE/ACS 14th International Conference on Computer Systems and Applications (AICCSA). https://doi.org/10.1109/aiccsa.2017.190.
10. Abdelkader Aissioui, Adlen Ksentini, Abdelhak Gueroui, and Tarik Taleb, "Towards Elastic Distributed SDN/NFV Controller for 5G Mobile Cloud Management Systems", IEEE Access, 2015, pp (1–10).
11. Pablo Jesus Argibay-Losada, Yuki Yoshida, Akihiro Maruta, and Ken-ichi Kitayama, "Optical Versus Electronic Packet Switching in Delay-Sensitive 5G Networks: Myths Versus Advantages", IEEE access, Nov 2016, Vol. 8, No. 11, pp. (43–54).
12. Miloud Bagaa, Tarik Taleb, Abdelquoddouss Laghrissi, Adlen Ksentini, and Hannu Flinck, "Coalitional Game for the Creation of Efficient Virtual Core Network Slices in 5G Mobile Systems", IEEE Journal on Selected Areas in Communications, 2018, pp (1–18).
13. Arsany Basta, Andreas Blenk, Klaus Hoffmanny, Hans Jochen Morpery, Marco Hoffmanny, Wolfgang Kellerer, "Towards a Cost Optimal Design for a 5G Mobile Core Network based on SDN and NFV", IEEE Transactions on Network and Service Management, 2017, pp (1–14).
14. Aydin Behnad, and Xianbin Wang, "Virtual Small Cells Formation in 5G Networks", IEEE Communications, 2016, pp (1–4).
15. Nikos Bizanis and Fernando Kuipers, "SDN and virtualization solutions for the Internet of Things: A survey," IEEE Access, 2016, pp (1–15).
16. Divya Chitimalla, Koteswararao Kondepu, Luca Valcarenghi, Massimo Tornatore, and Biswanath Mukherjee, "5G Fronthaul–Latency and Jitter Studies of CPRI Over Ethernet", IEEE-Journal. Opt. communication. Network, Feb 2017, Vol. 9, No. 2, pp. (172–182).
17. Gabriel Otero Pérez, José Alberto Hernández, and David Larrabeiti, "Fronthaul Network Modeling and Dimensioning Meeting Ultra-Low Latency Requirements for 5G", IEEE, JUNE 2018 VOL. 10, NO. 6, pp (573–581).
18. Sadia Din, Anand Paul, Awais Ahmad, Syed Hassan Ahmed, Gwanggil Jeon, Danda B. Rawat, "Hierarchical Architecture for 5G based Software-Defined Intelligent Transportation System", IEEE Conference on Computer Communications, 2018, pp (462–467).
19. Dahou Djilali, Y.A Bakhti, Y. Kouninef, B., Senouci, B. "Performances Evaluation study of VANET Communication Technologies for Smart and Autonomous Vehicles," IEEE 2018, pp (79–84).
20. Velliangiri, S. (2020). An Enhanced Multimedia Video Surveillance Security Using Wavelet Encryption Framework. Journal of Mobile Multimedia, 15 (3), 239–254.
21. Jiaxin Zhang, Xing Zhang, Muhammad Ali Imran, Barry Evans, Yan Zhang, and Wenbo Wang, "Energy Efficient Hybrid Satellite Terrestrial 5G Networks with Software Defined Features", Journal Of Communications And Networks, Apr 2017,Vol. 19, No. 2, pp (147–162).
22. Matteo Fiorani, Ahmad Rostami, Lena Wosinska, and Paolo Monti, "Abstraction Models for Optical 5G Transport Networks", IEEE-J. Opt. Communication. Network, Sept 2016, Vol. 8, NO. 9, pp. (656–665).
23. Velliangiri, S., & Premalata, J. (2020). A Novel Forgery Detection in Image Frames of the Videos Using Enhanced Convolutional Neural Network in Face Images. Computer Modeling in Engineering & Sciences, 125(2), 625–645.

24. Xiaohu Ge, Junliang Ye, Yang Yang, Qiang Li, "User Mobility Evaluation for 5G Small Cell Networks Based on Individual Mobility Model", IEEE Journal on Selected Areas in Communications, 2016, pp. (1–15).
25. Wanqing Guanyz, Xiangming Wenyz, Luhan Wangyz, Zhaoming Luyz, Yidi Shenyz, "A Service-oriented Deployment Policy of End-to-End Network Slicing Based on Complex Network Theory," IEEE. Translations and content mining, 2018, pp. (1–11).
26. Akhil Gupta, Rakesh Kumar Jha, "A Survey of 5G Network: Architecture and Emerging Technologies," IEEE Access, pp. (1–26).
27. Mahmoud A. Hasabelnaby, Hossam A. I. Selmy, and Moawad I. Dessouky, "Joint Optimal Transceiver Placement and Resource Allocation Schemes for Redirected Cooperative Hybrid FSO/mmW 5G FronthaulNetworks", IEEE journal, Dec 2018, Vol. 10, No. 12, pp (975–990).
28. Shao-Chou Hung, Hsiang Hsu, Shao-Yu Lien, and Kwang-Cheng Chen, "Architecture Harmonization between Cloud Radio Access Networks and Fog Networks," IEEE Access, Dec 2015, pp. (3019–3034).
29. P. Karthikeyyan, S. Velliangiri and M. I. T. Joseph. S., "Review of Blockchain based IoT application and its security issues," 2019 2nd International Conference on Intelligent Computing, Instrumentation and Control Technologies (ICICICT), Kannur, India, 2019, pp. 6–11, https://doi.org/10.1109/ICICICT46008.2019.8993124.
30. https://www.researchgate.net/figure/VANET-Cloud-Architecture_fig1_290447374

Chapter 11
Remote Sleep Monitoring and 5G

B. L. Radhakrishnan, E. Kirubakaran, V. Ebenezer, R. V. Belfin, and Derrick I-Hsien Ting

11.1 Introduction

Sleep is an essential part of human life, and about one-third of our time is spent on sleeping [1]. According to National Sleep Foundation recommendations, different age groups need distinct sleep duration, school children (6–13 years): 9–11 h, teenagers (14–17 years): 8–10 h, young adults and adults (18–64 years): 7–9 h, and older adults (65+ years): 7–8 h [2]. Getting enough sleep (quality sleep) at the right time is essential for survival, and lack of sleep leads to physical and cognitive dysfunction [3]. Conditions that affect sleep timing, sleep quality, or sleep duration are referred to as sleep disorder; it impacts humans' normal functioning when they are awake and can even lead to medical issues [4]. Research indicates that the risk of diseases like high blood pressure [5], cardiovascular disease [6], diabetes [7], depression [8], and obesity [9] is enhanced by a prolonged lack of sleep or low-quality sleep. The golden standard to measure sleep is polysomnography. This method uses a various surface electrode to measure physiological sleep parameters, such as electroencephalography (EEG), eye movements, muscle movement, heart rate and respiratory rates. This method's disadvantage is that individuals have to spend the night at a sleep clinic, under a controlled setup and continuous monitoring of sleep technician [10].

B. L. Radhakrishnan (✉) · E. Kirubakaran · V. Ebenezer · R. V. Belfin
Department of Computer Science and Engineering, Karunya Institute of Technology and Sciences, Coimbatore, India
e-mail: blradhakrishnan@karunya.edu; ekirubakaran@karunya.edu; ebenezerv@karunya.edu

D. I.-H. Ting
Department of Information Management, National University of Kaohsiung, Kaohsiung, Taiwan
e-mail: iting@go.nuk.edu.tw

S. Velliangiri et al. (eds.), *Secure Communication for 5G and IoT Networks*, EAI/Springer Innovations in Communication and Computing,
https://doi.org/10.1007/978-3-030-79766-9_11

Health monitoring in real-life conditions (out-of-hospital conditions) has been of constant interest to researchers and healthcare professionals [11]. One of the significant advantages of the digital era is remote health monitoring; it enables tracking or monitoring an individual's health-related parameters remotely and share with the healthcare providers [12]. Recent developments in sensors, microcontrollers, and communication systems play an important role in remote healthcare; IoT, wearable devices, nearable devices and smartphones use sensors to acquire health data from individuals [13]. The Internet of Things (IoT) has driven the healthcare sector to use entirely digitized e-health systems and plays an inevitable role in remote healthcare [13]. The application of wearable sensors to assess an individual's health or well-being status is most attractive in out-of-hospital circumstances [11]. Smartphones became a fundamental need of modern society, including the medical field; among the public, there has been growing interest in using apps to improve fitness and wellness [14]. The adoption of artificial intelligence (AI) in remote healthcare improves patient care. AI plays a vital role in healthcare areas such as automated diagnosis, information analysis, personalization, coordinated care, decision-making, virtual interactions, continuous monitoring and alerting, skill-based doctor matching, and so on [15].

Sleep disorders have a high prevalence in the general population: insomnia in adults (10–20%), and sleep apnea (4–6%). Usually, sleep investigations are carried out in sleep labs; the procedure is costly, and individuals have to wait for a long time. An alternative to sleep lab studies could be remote sleep monitoring [16]. IoT [17], wearables and nearables [18] and smartphone applications [14, 19] widely used in tracking sleep issues. In remote sleep monitoring, IoT devices carry out sleep monitoring [20], sleep quality assessment or monitoring [21, 22], sleep apnea detection [23], sleep quality assessment [24], and sleep apnea treatment support and diagnosis [25, 26]. Besides, in remote sleep monitoring, wearable and nearable devices are used to perform contactless sleep monitoring [27], sleep signal monitoring [28], sleep assessment [29], sleep apnea detection [30], sleep quality monitoring [31], and detect cardiovascular event based on sleep [32]. Similarly, the smartphone-based applications are used to track sleep apnea screening [33, 34], detecting hypopneas and apneas [35], Cognitive behavioral therapy for insomnia patients [36], detecting sleep–wake-up cycles [37], sleep monitoring [38], identifying sleep patterns [39], sleep onset estimation [40], snoring detection [41], sleep behavior disorder [42], and sleep stages [43].

Remote sleep monitoring systems transfer the sensor's data to the remote server or cloud; it highly relies on existing communication technologies for data transfer. Sleep has an association with physical and mental health. Continuous monitoring of sleep enables the healthcare systems to detect deteriorating physical and mental in advance [44]. Table 11.1 shows the different communication techniques used in the remote healthcare system.

Cities are transitioning into smart cities and adopting remote health monitoring on the rise that significantly increase the number of devices connected to the network. This transformation succeeds if there is the capacity to support every IoT

Table 11.1 Communication techniques used in the remote healthcare system

Communication protocols	Frequencies	Range	Data rates	Power consumption
Bluetooth 4.2	2.4 GHz	50–150 m	1 Mbps	Low
Bluetooth 5	2.4 GHz	0.25 km	2 Mbps	Very low
RFID	13.56 MHz–2.45 GHz	1–100 m	40–640 kbps	Very low
Zigbee 3.0	2.4 GHz	10–100 m	250 kbps	Very low
Z wave	Various	30 m	0.3–50 kbps	Very low
NFC	13.56 MHz (ISM)	10 cm	100–420 kbps	Very low
LoRaWAN	Various	2.5 km (UE), 15 km (SE)	0.3–50 kbps	Low
Sigfox	900 MHz	30–50 km (RE), 3–10 km (UE)	10–1000 bps	Low
GSM/GPRS	850, 900, 1800, 1900 MHz	5–30 km	80–384 kbps	High
3G	850 MHz	5–30 km	3 Mbps	High
4G	700, 1700, 2800 MHz	10 km max	Up to 12 Mbps	High
5G	Low band/high band	10 km max	3.6 Gbps max/10 Gbps	High
Wifi	2.4–5 GHz	50–100 m	150–200 Mbps	Medium

UE Urban environment, *SE* Suburban environment, *RE* Rural environments

and other connected devices' connectivity [44]. The remote health monitoring systems use any available communication protocols listed in Table 11.1.

The remainder of this chapter is organized as follows. Section 11.2 describes remote sleep monitoring. Section 11.2 comprises of three sub-sections; Sect. 11.2.1 IoT-based remote sleep monitoring, Sect. 11.2.2 Wearable and nearable based sleep monitoring, and Sect. 11.2.3 Smartphone application-based sleep monitoring. Section 11.3 discusses the role of 5G in remote sleep monitoring. Section 11.4 presents future directions, focusing on the randomized control trials among multiethnic groups and blockchain-based frameworks to improve end user device security and integrity.

11.2 Remote Sleep Monitoring

Remote sleep monitoring is an emerging modern sleep monitoring model, facilitating real-time monitoring and sleep management of individuals from home. Due to the COVID-19 pandemic, people's lifestyle and work nature have significantly changed and contributed to people's physical and mental health illness. Better integration of remote sleep monitoring system helps us to predict the psychological and physical adversities early. We classify the remote sleep monitoring based on the monitoring methods. Figure 11.1 shows the major classification of sleep monitoring techniques.

11.2.1 IoT-Based Remote Sleep Monitoring

In remote healthcare, IoT plays an indispensable role in integrating patient, hospital and medical equipment [25]. In standard polysomnography assessment, the patient has to visit the sleep lab or hospital, and a trained sleep technician should present to carry out the sleep study. Besides, long waiting queues in sleep clinics, discomfort due to sensors' placement, and routine sleep place change are common issues with polysomnography [20]. In the assessment of sleep, remote sleep monitoring using IoT provides similar diagnostic details to laboratory polysomnography in evaluating sleep and related disorders [20]. Sleep is very important for human mental and physical health. Ample sleep can reduce mental depression and assist the brain to work smartly. A reliable monitoring system is needed to monitor the sleep quality of an individual.

IoT [45] technology plays a major role to build a sleep monitoring system. IoT interconnects device, people and anything which is connected to the internet [46]. The sleep monitoring system monitors the sleep quality of the human being. The architecture for sleep monitoring system is shown in Fig. 11.2. Data Acquisitions comprises the data acquired through wireless sensors [47], and movement detection service. Data aggregator [48] comprises a smartphone/cloudlet service used to collect and forward the sleep data to cloud storage. Cloud storage and processing [49] consist of public or private cloud, which is used to access larger and faster data for sleep monitoring.

The doctor or concerned person will review the sleep result. The sensor data are streamed to the storage. The sensors supply respiration rate, heart rate and snoring data, and so on. Table 11.2 shows the progress of IoT-based systems for sleep monitoring, diagnosis, and treatment models.

The IoT-based remote sleep system studies listed in Table 11.2 show that these systems use sensors such as PIR motion, accelerometer (integrated/semifixed), accelerometer, microphone, pulse oximeter, EEG, EMG, EOG, and ECG,

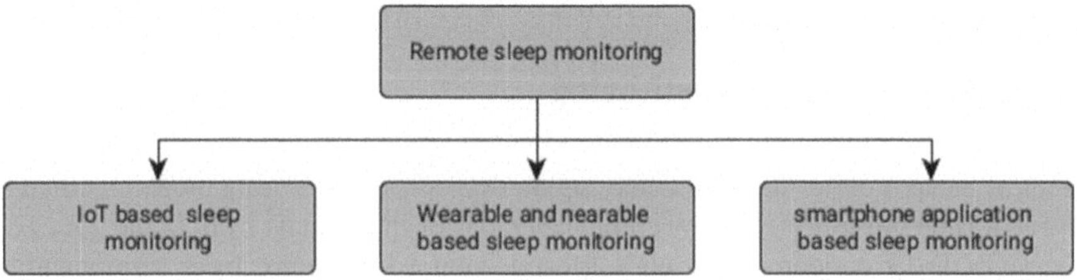

Fig. 11.1 Remote sleep monitoring types

Table 11.2 IoT-based systems for sleep monitoring, diagnosis, and treatment models

Author, year	Study objective	Sleep monitoring type	Sensors/devices/tools	Platform	Results validation	Outcome
Kim, Chu and Kang [50]	Sleep–wake detection in elderly	Bed based	PIR motion sensors, accelerometer sensor (semi fixed)	Android	PSQI	Sensitivity (96.61%), specificity (91.81%), AUC performance (94.21%.)
Chang et al. [51]	Sleep estimation in in insomnia, and euthymic patients	Bed based	Integrated sensor	Android	NA	Sensitivity (65%), specificity (74%), accuracy (84%)
Saleem et al. [22]	Analyze sleep patterns	Movement based	Accelerometer, microphone, pulse oximeter	Android	Test datasets	Sensitivity (65%), specificity (74%), accuracy (95%)
Mendonça et al. [23]	OSA diagnosis using SpO_2	Bed based	ECG and SpO_2 sensor	Android	PSG	Sensitivity (80%), specificity (91%), accuracy (95%)
Ma et al. [52]	OSA diagnosis using SpO_2	Movement based	SpO_2, temperature and humidity sensors	Android	PSG	Sensitivity (87.6%), specificity (94.1%), accuracy (90.2%)
Azimi et al. [21]	Assess maternal sleep quality	Movement based	EEG, EMG, EOG, and ECG	Android	PSG	Predict sleep abnormalities during pregnancy and postpartum
Yacchirema et al. [53]	Measure sleep apnea using wearable and IoT	Movement based	SpO_2, temperature and humidity, steps counter sensors, wearable wristband, BMI	Android	NA	Classified no apnea; mild, moderate, and severe categories
Yacchirema et al. [26]	OSA detection and treatment support	Autonomic based	Environmental sensor and motion sensor	Android	Test data sets	Air quality prediction efficiency (93.3%) and guide OSA treatment
Wu et al. [25]	OSAHS diagnosis and treatment	Movement based	ECG, WRSBM	Android	Test datasets	Improvement in before and after treatment ($p < 0.05$)
Lin et al. [20]	Sleep monitoring using wireless polysomnography	Bed based	EEG, EOG, EMG, ECG, and airflow sensor	Android	PSG	Better sleep stage classification, good for OSA detection

PIR Passive infrared IoT-based sleep, *BMI* body mass index, *WRSBM* wireless remote sleep breathing monitor, *OSA* obstructive sleep apnea, *OSAHS* obstructive sleep apnea/hypopnea syndrome

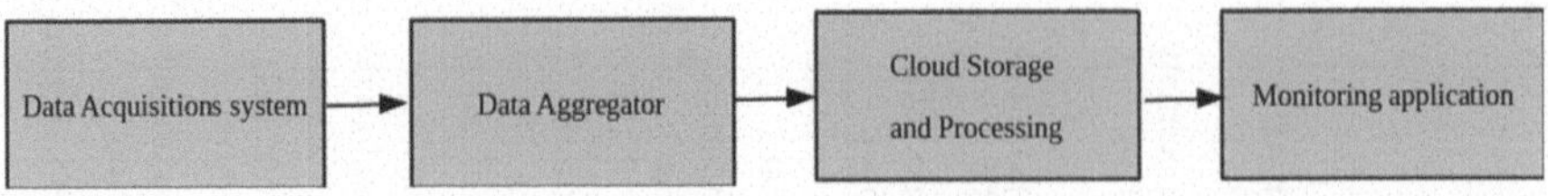

Fig. 11.2 IoT architecture for sleep monitoring

temperature, humidity, step counter, environmental, motion, WRSBM, and airflow to track sleep parameters for better performance or validation.

The IoT-based remote sleep monitoring system listed in Table 11.2 shows a remarkable improvement in tracking sleep parameters over the years. Some of the systems achieved more than 90% sensitivity, 95% accuracy, and greater than 90% specificity. Some of the systems show excellent results that are on par with PSG and PSQI. However, the observed outcomes were not significant for clinical purposes since the testings were done on fewer people who were either healthy or patients. More validated studies need to be done among people with different ethnic, cultural, and age groups. A more elevated focus on incorporating artificial intelligence could produce more accurate results.

11.2.2 Wearable and Nearable Based Sleep Monitoring

The commercial market for technologies to monitor and promote sleep is expanding at a faster pace. A wide range of smartwatches, bands, garments, and patches with embedded sensors, small portable devices and mobile applications now exist to record and provide users with feedback on their sleep. The work by Peake, Kerr and Sullivan [54] classified the sleep monitoring devices and discussed the research validation and device reliability testing on these devices. Figure 11.3 shows types of wearable and nearable devices used in remote sleep monitoring.

Peake et al. classified devices like wearables and nonwearable. In this section, the devices (wearables and nearables) have been classified based on how the users use them. Table 11.3 shows the details about sleep monitoring and promoting products available in the market with their technical information.

In this section, we have done a clear literature survey on the devices used for the sleep study. Fitbit has been used in research to collect the sleep parameter like sleep quality, fitness level, heart rate, and breathing pattern. Many studies [71–74] attempted to evaluate the use of the Fitbit band to collect sleep data. OURA also has been used for many sleep studies. The works like [75–77] used OURA ring for Sleep quality studies, Sleep tracker studies and sleep environment-related studies. The Dreem headband has been used as the alternative for polysomnography in many studies [78–82]. Dreem headband device has been mostly used for sleep staging-related studies. Sleep Profiler is a headband connected to a web application and helps to track the pulse rate, head position, head movement, sleep latency and many more. There are few studies [83, 84] used this sleep profiler to monitor sleep,

Table 11.3 Sleep monitoring and promoting products

Product name	Product description	Product category	Technical outcome	Sensors	Validated
FitBit [55]	Tracks sleep, exercise, food, and weight	Wristband, wrist watches	Resting heart rate, cardio fitness level, guided breathing, heart rate zones	PurePulse	Yes
OURA [56]	A ring that combines sensor and minimal design to provide precise, personalized health insights	Finger ring + Mobile app	Resting heart rate, Heart rate, variability (HRV), Respiratory rate Body, temperature Light, deep and REM sleep, Night-time movement, Sleep timing and quality	PPG sensors NTC temperature Sensors, Accelerometer	Yes
Dreem [57]	Monitoring and automatic analysis of EEG and biosignals	Head band + Mobile app	Heart rate, reports of sleep endpoints(TST, WASO, N1, N2, N3, REM Sleep, SOL, nb, and duration of awakening)	6 EEG sensors, Pulse sensor, Accelerometer	Yes
Sleep Profiler [58]	Web-based presentation of recorded signals with validated auto-staging and technical editing capabilities	Head band + web app	Pulse rate, head position, head movement, quantitative, snoring, sleep latency, REM latencoimbatorecy, N3 latency	EEG, EOG, EMG, EMG or ECG	Yes
Somté PSG [59]	Sleep and breathing record device	Skin sensors + thoracic band + abdominal band	SpO_2, pulse rate, plethysmography waveform and oximeter probe status	EEG EOG Chin EMG ECG—with thoracic impedance option	Yes
Zmachine—Synergy [60]	EEG-based sleep staging technology with traditional respiratory measures	EEG skin sensors + Zmachine system + chest strap	Advanced sleep reports (AHI values, RERA)	EEG, 3D body position sensor	No

(continued)

Table 11.3 (continued)

Product name	Product description	Product category	Technical outcome	Sensors	Validated
Zmachine Insight + [61]	EEG-based sleep staging	EEG skin sensors + Zmachine system (wall hanging)	Raw EEG data + data analytics + dashboard for statistics	EEG	No
Sleep Shepherd [62]	Sleep tracker with a feedback system	Head band + Mobile app	Brain rate related to sleep	EEG, binaural beat system	No
Re-Timer [63]	Solution to improve sleep using green–blue light therapy	Eye wear	Improves sleep	Light therapy using green and blue LED lights	Yes
AYO [64]	Solution for better sleep using light therapy	Eye wear	Improves sleep	Capacitive sensor, ambient light sensor	No
HUSH [65]	Hush enables the user to sleep well	Ear plug	Improve sleep, filters unwanted noise	Noise masking, Tranquilities, Naturescapes	No
Kokoon [66]	Personalized audio and coaching for better sleep	Head phone + Mobile app	Track sleep and give personalized recommendations	EEG brainwave sensors, Motion sensing, Disturbance sensing	No
Dreampad [67]	Dreampad is a drug-free solution to better sleep, using music	Pillow + app	Can reduce stress and anxiety and help to get good sleep	Vibrational sound system	No
Beddit Sleep Monitor [68]	Automatic and accurate tracking of sleep during night	Thin sensor strip placed under the bed sheet	Sleep tracking, heart rate, breathing, snoring, environment	Piezo force, Capacitive touch, Humidity, Temperature	Yes
CPAP Equipment [69]	(CPAP) machine is used to treat sleep apnea	CPAP machine + CPAP MASK + Mobile APP	Help to get better sleep for sleep apnea patients	NA	No

(continued)

Table 11.3 (continued)

Product name	Product description	Product category	Technical outcome	Sensors	Validated
HexoSkin [70]	Comfortable and washable garment to monitor your health, activity, and sleep	Smart garment + Mobile APP	HRV (for stress monitoring, effort, load, and fatigue assessments), QRS events, heart rate zones, heart rate maximum, resting heart rate, and heart rate recovery	RIP sensors ECG	No

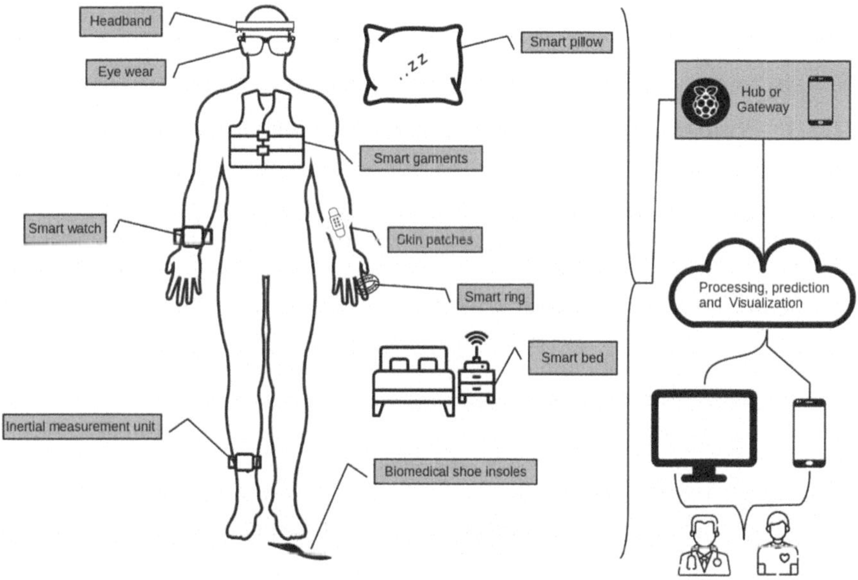

Fig. 11.3 Types of wearable and nearable devices used in remote sleep monitoring

Identify the Sleep EEG Biomarker and head position monitoring during sleep for patients with neurodegenerative disease. Somté PSG has been used in many studies [83, 84] to test, diagnose and monitor the sleep apnea. There are few applications which used this device for telehealth supported patient monitoring. Zmachine is an EEG-based skin patches used to acquire the sleep parameters like sleep duration, sleep patterns, and sleep quality. The studies [72, 85] used this device for deriving the sleep parameter and applications like sleep–awake detection system and sleep monitoring systems. Re-Timer is a technology of light therapy by using green and blue light to improve sleep. Studies like [74, 85, 86] used the Re-Timer eyewear for their research. The applications like sleep measurement, sleep monitoring, sleep extension, and mood promotion to sleep have been done using this device.

Dreampad is a nearable pillow that can play the music the user likes and help them sleep better. Studies like Wolfhope and Hudkins [87], Gutman et al. [88], Schoen, Man, and Spiro [89], Srikolcheep and Sittiprapaporn [90], Jones and Moyle [91] used the Dreampad for the sleep studies. This device has been used in applications like sleep quality among children with autism, improving sleep quality among insomnia patients, and many other occupational therapy applications. Beddit sleep monitor is a nearable product developed from Apple Inc for sleep studies. The studies [88–91] have been focused on sleep monitoring and telerehabilitation application. CPAP equipment improves the breathing, which in turn improves sleep. The studies like Lain and Bourn [92], Mishra et al. [93], Leger et al. [94] worked with this device for their sleep and respiratory distress-related research. HexoSkin is a wearable garment that can monitor health, sleep, and activity. The work by PSFK [95] and Slamon et al. [96] explored various research with the smart garment related to environment for sleep and other sleep-related monitoring and tracking studies.

11.2.3 Smartphone Application-Based Sleep Monitoring

The expeditious adoption of smartphones by the general public and healthcare sector professionals has created many opportunities in remote healthcare sectors. Smartphones play an inevitable role in remote sleep management, monitoring and tracking. In sleep monitoring, there has been an increased interest in remote sleep reflected by both increasing research works in this area and diverse consumer apps targeting remote sleep monitoring [19]. In Apple's app store and Google's play store, the number of sleep-related apps continues to grow every year [14]. Sleep apps provide features to detect the sleep duration, awake state, light sleep, deep sleep, REM sleep, sleep efficiency and sleep debt. Some apps offer additional features such as motion tracking, snore recording, smart alarm, sleep assist, notes, and heart rate monitoring [14]. Sleep monitoring or tracking apps have the potential to increase awareness and good sleep habits. Nevertheless, without clinical validation and compliance, these apps cannot be used by patients or healthy people for clinical remote monitoring [19]. Table 11.4 shows smartphone apps validation studies to assess sleep monitoring, sleep apnea detection, monitoring, and sleep pattern.

The smartphone remote sleep app validation studies listed in Table 11.4 show that sleep apps use inbuilt smartphone sensors such as microphones, light sensor, accelerometer, gyroscope, and camera for sleep parameters. Moreover, sleep apps use external sensors or devices like Fitbit, three-axis accelerometer, earbud, external microphones, pulse oximeter to track sleep parameters for better performance, or validation.

The smartphone remote sleep apps listed in Table 11.4 show a phenomenal improvement in tracking sleep parameters over the years. Some of the apps achieved 90% sensitivity, greater than 80% accuracy, and nearly 90% specificity. Some of the sleep apps are approximately providing on par results of PSG, actigraphy, and

Table 11.4 Smartphone app-based sleep monitoring, screening, and diagnosis

Author, year	Study objective	Country	No of participants	Health status	Device placement	Platform	Sensor/device / tools	Results validation	Outcome
Ferrer-Lluis et al. [97]	Design an algorithm to detect OSA	Spain	13	OSA patients	Over sternum	Android	MPU-6500	Apnealink	Estimating AHI sensitivity = 90%; predictive value = 80%
Lee, Marek and Nálevka [98]	Sleep pattern during COVID	China	563	Healthy	On the bed	Android	External wearable (fitbit), built-in accelerometer and light sensor	NA	Increased SD = μ 20 min, DSO = μ 30 min
Philip et al. [99]	Monitor insomnia symptoms and acceptance of app during COVID	France	2069	Healthy and insomnia patients	On the bed	Android	AI-based virtual agent	ISI	Acceptance (61.7%), usability (91.6%)
Tiron et al. [100]	OSA screening	Germany	248	Healthy individuals	Above the mattress height, charger port pointing toward the chest	Android, IOS	Built in microphone, speaker, gyroscope, and accelerometer; firefly AI/ML algorithms	Test datasets	Sensitivity (88.3%), specificity (80.0%)
Ren et al. [38]	Sleep monitoring	NA	9	Healthy	Right side of bed	IOS	Earphone, built in accelerometer,	NEULOG	Better detection of BR, sleep events, and movement; good for OSA monitoring

Author, year	Study objective	Country	No of participants	Health status	Device placement	Platform	Sensor/device / tools	Results validation	Outcome
Narayan et al. [101]	OSA diagnosis	USA	91	Individuals with sleep disorders	On the bedside	Android	Built-in microphone	PSG	Acoustic signal analysis c-statistic 0.87
Castillo-Escario et al. [35]	OSA diagnosis	NA	13	Healthy and sleep apnea patients	On the chest	Android	Built-in microphone and accelerometer	ApneaLink	Sensitivity (76%), positive prediction (82%), accuracy (82%)
Borger, Huber and Ghosh [37]	Measure sleep–wake cycles	Netherlands	79	Healthy	NA	Android	Tappigraphy	Actigraphy	$R2 = 0.9$, $\beta = 0.90$, $t(1380) = 111.4$, $p < 0.0001$)
Scott, Lack and Lovato [40]	Sleep onset estimation	Australia	12	Healthy	Strapped on the dominant forearm	IOS	Earbud	PSG	$r = 0.79$, $P < 0.001$
Patel, Kim and Brooks [102]	Sleep estimation	USA	25	OSA patients	Near the pillow	IOS	Built-in accelerometer and microphone	PSG	SL of app and PSG ($p = 0.03$), sleep stages no correlation
Gu et al. [43]	Sleep stage tracking	NA	15	Healthy	Near the pillow	Android	Built-in microphone, accelerometer and light sensor	Jawbone, Zeo	Detects sleep stages better than jawbone and Zeo

(continued)

Table 11.4 (continued)

Author, year	Study objective	Country	No of participants	Health status	Device placement	Platform	Sensor/device / tools	Results validation	Outcome
Behar et al. [33]	OSA screening	UK	856	OSA patients	Strapped using armband	Android	Built-in microphone and accelerometer; external microphone and wrist worn pulse oximeter	Database of Churchill Hospital, Oxford	Audio, actigraphy accuracy (88.4%), ODI accuracy (92.3%)
Bhat et al. [103]	Sleep monitoring	USA	20	Healthy	Placed on the bed	IOS	Built-in accelerometer and microphone	PSG	No correlation between app and PSG SE ($r = -0.127$, $p = 0.592$), accuracy (45.9%), sleep–wake; accuracy (85.9%), sensitivity (89.9%), specificity (50%)
Nakano et al. [104]	OSA detection	Japan	50	OSA patients	Anterior chest wall, microphone pointing toward the neck	Android	Built-in microphone	PSG	Snoring time highly correlated ($r = 0.93$), RDI highly correlated with AHI ($r = 0.94$), diagnostic sensitivity (0.70) and specificity (0.94)

OSA obstructive sleep apnea, *PAP* positive airway pressure, *ODI* oxygen desaturation index, *RDI* respiratory disturbance index, *AHI* apnea–hypopnea index, *SD* sleep duration, *DSO* delayed sleep onset, *SL* sleep latency, *SE* sleep efficiency, *BR* breathing rate, μ mean, *NA* not applicable

ApneaLink. However, the observed outcomes of the sleep apps in this study were not significant for clinical purposes since the testings were done on fewer people of either healthy or patients. More validation studies need to be done among people with different ethnic, cultural, and age groups. Artificial intelligence-based approaches in sleep apps have significantly improved usability and acceptability [99]. A greater focus on smartphone sleep app algorithms and incorporating artificial intelligence could produce more accurate results.

11.3 Role of 5G in Remote Sleep Monitoring

Remote health monitoring or sleep monitoring systems, especially IoT-based systems, heavily rely on communication networks such as mobile networks (2G, 3G, 4G), Wi-Fi, Bluetooth, and satellite communication to transferring data to the remote network or cloud. However, the networks mentioned above cannot support connectivity to high mobility devices, connectivity transition, and connectivity to many devices [44]. The next-generation 5G networks play a significant role in remote sleep monitoring systems because of their potential properties such as increased accessibility, improved mobile broadband reach, and increased health and security. 5G can play a major role in meeting the present requirements of communication in remote sleep monitoring systems, with the following features [105]

- Low latency and high throughput
- Increased network capacity
- Enhances the performance of edge computing
- Connectivity to all devices (IoT, wearable, nearable, and smartphone)

The advantage of low latency and high throughput is enhanced traffic capacity and high data rate. In [106], the author mentioned the following requirements; suburban and urban areas data rate is 100 Mbps and 1 bps in hot spots; the maximum data rate is 20 Gbps. Increased network capacity is related to the machine to machine (M2M) communication. A recent investigation by [107] discussed the connection density of 106 devices/m^2 in remote healthcare and edge computing and energy efficiency 100 times improved when compared with 4G.

Traditionally, in remote sleep monitoring settings, the data is sent to the remote computer or cloud for processing that introduced latency in getting the results, heavy bandwidth usage and privacy concerns. There might be a requirement for processing and analyzing the data in the local environment in some scenarios. However, running AI applications in resource-constrained environments is not a viable option. Edge computing is a type of distributed computing model that brings the computing and storage to the network's edge to maximize the computing experience. 5G-powered edge computing used in resource-constrained settings to reduces the latency in producing results that enable it to perform more complex tasks [108, 109].

For example, in global healthcare, IoT wearable and nearable markets are projected to touch 158.1 billion USD in 2022 [110]. In order to enable widespread IoT, wearable and nearable adoption, 5G networks will play a significant role [111].

According to the statistics facts and prediction, in 2025, there will be 75 billion IoT devices [112], 520.1 million wearable and nearable devices [113], and 2779 million 5G-enabled smartphones connected to the internet [114]. The design of 5G is to accommodate billions of connected devices and increased data growth to meet the need of modern society. It is expected to take the remote sleep monitoring applications to the next level [115].

11.4 Future Directions

Remote sleep monitoring is gaining much attention because of its flexibility compared with the traditional polysomnography test in a traditional sleep clinic. The existing remote sleep monitoring studies discussed in this chapter achieved 90% or higher accuracy, sensitivity, and specificity levels. Even though remote sleep monitoring studies give good results, it is necessary to get the following. First, more accurate algorithms should be developed to detect sleep disorders and classifying sleep stages using remote monitoring systems near real time. Second, there is a concern about the validation studies and the Food and Drug Administration (FDA) clearance regarding the remote sleep monitoring systems, particularly wearable devices [116]. Hence, researchers can improve remote sleep monitoring systems by conducting more evidence-based studies using randomized control trials among multiethnic groups. Although there are limitations in the remote sleep monitoring systems, they can be instrumental in situations like COVID-19 lockdown to monitor sleep patients [117].

5G is going to play a significant role in remote sleep monitoring systems. Modern healthcare systems depend on cloud computing for their computation and data storage needs due to resource requirements. In the emerging 5G-based remote sleep monitoring models, the following problems would arise: first, end-user devices are more vulnerable to attack due to a lack of software-based measures and configuration. Second, there are no user integrity protection mechanisms [118]. More framework-based approaches are essential for remote sleep monitoring systems using blockchain to improve the security in end-user devices and data and user integrity [115].

11.5 Conclusion

Prevalence of sleep or sleep-related disorders is common in modern society. Remote sleep monitoring systems seem very promising for smart home and smart cities that reduce the burden of traditional sleep monitoring; it supports continuous sleep monitoring, which is impossible in conventional settings. According to various research studies, sleep has an indispensable role in physical and mental health issues. This chapter's discussions suggest that advancements in sensors and AI application have improved remote healthcare to the next level. Most of the evaluation studies discussed in this chapter lack proper validation done on different age and ethnic groups. Hence, proper validation studies need to be carried out to improve the reliability of these systems. Additionally, remote sleep monitoring suffers from communication issues and resource constraints. Implementation of 5G in remote sleep monitoring systems will improve user experience, prediction accuracy, advance care, better processing speed, and greater bandwidth for data transfer.

References

1. Held, G. (2017) 'Understanding Sleep', *www.ninds.nih.gov*, NIH Public(NIH Publication No. 17-3440c).
2. Hirshkowitz, M., Whiton, K., Albert, S. M., Alessi, C., Bruni, O., DonCarlos, L., Hazen, N., Herman, J., Katz, E. S., Kheirandish-Gozal, L., Neubauer, D. N., O'Donnell, A. E., Ohayon, M., Peever, J., Rawding, R., Sachdeva, R. C., Setters, B., Vitiello, M. V, Ware, J. C. and Adams Hillard, P. J. (2015) 'National Sleep Foundation's sleep time duration recommendations: methodology and results summary.', *Sleep health*. National Sleep Foundation, 1(1), pp. 40–43. https://doi.org/10.1016/j.sleh.2014.12.010.
3. Krueger, J. M. and Obal, F. (2003) 'Sleep function', *Frontiers in Bioscience*, 8(3), pp. 511–519. https://doi.org/10.2741/1031.
4. Chokroverty, S. (2010) 'Overview of sleep & sleep disorders', *Indian Journal of Medical Research*, 131(2), pp. 126–140.
5. Knutson, K. L., Van Cauter, E., Rathouz, P. J., Yan, L. L., Hulley, S. B., Liu, K. and Lauderdale, D. S. (2009) 'Association between sleep and blood pressure in midlife: The CARDIA sleep study', *Archives of Internal Medicine*, 169(11), pp. 1055–1061. https://doi.org/10.1001/archinternmed.2009.119.
6. Hoevenaar-Blom, M. P., Spijkerman, A. M. W., Kromhout, D., Van Den Berg, J. F. and Verschuren, W. M. M. (2011) 'Sleep duration and sleep quality in relation to 12-year cardiovascular disease incidence: The MORGEN study', *Sleep*, 34(11), pp. 1487–1492. https://doi.org/10.5665/sleep.1382.
7. Knutson, K. L., Ryden, A. M., Mander, B. A. and Van Cauter, E. (2006) 'Role of sleep duration and quality in the risk and severity of type 2 diabetes mellitus', *Archives of Internal Medicine*, 166(16), pp. 1768–1774. https://doi.org/10.1001/archinte.166.16.1768.
8. O'Leary, K., Bylsma, L. M. and Rottenberg, J. (2017) 'Why might poor sleep quality lead to depression? A role for emotion regulation', *Cognition and Emotion*, 31(8), pp. 1698–1706. https://doi.org/10.1080/02699931.2016.1247035.
9. Rahe, C., Czira, M. E., Teismann, H. and Berger, K. (2015) 'Associations between poor sleep quality and different measures of obesity', *Sleep Medicine*, 16(10), pp. 1225–1228. https://doi.org/10.1016/j.sleep.2015.05.023.
10. Jafari, B. and Mohsenin, V. (2010) 'Polysomnography', *Clinics in Chest Medicine*, 31(2), pp. 287–297. https://doi.org/10.1016/j.ccm.2010.02.005.
11. Korhonen, I., Pärkkä, J. and Van Gils, M. (2003) 'Health Monitoring in the Home of the Future', *IEEE Engineering in Medicine and Biology Magazine*, 22(3), pp. 66–73. https://doi.org/10.1109/MEMB.2003.1213628.

12. Joaquin Delgado, S. F. S. (1994) 'MULTI-USER REMOTE HEALTH MONITORING SYSTEM', *United States Patent*, 54(19), pp. 3–5. Available at: https://patentimages.storage.googleapis.com/84/e6/5f/65765dbc491a4f/US5347263.pdf.
13. Eshaghi, S., Ghaziani, R. K. and Ansari, A. (2019) 'Sensors for Internet of Medical Things: State-of-the-art, Security and Privacy Issues, Challenges and Future Directions', *Mathematics and Computers in Simulation*, pp. 1–48. Available at: https://doi.org/10.1016/j.matcom.2019.11.009.
14. Ong, A. A. and Gillespie, M. B. (2016) 'Overview of smartphone applications for sleep analysis', *World Journal of Otorhinolaryngology-Head and Neck Surgery*. Elsevier Ltd, 2(1), pp. 45–49. https://doi.org/10.1016/j.wjorl.2016.02.001.
15. Kuziemsky, C., Maeder, A. J., John, O., Gogia, S. B., Basu, A., Meher, S. and Ito, M. (2019) 'Role of Artificial Intelligence within the Telehealth Domain Official 2019 Yearbook Contribution by the members of IMIA Telehealth Working Group', *IMIA Yearbook of Medical Informatics 2019*, (Official 2019 Yearbook Contribution by the members of IMIA Telehealth Working Group), pp. 35–40. Available at: https://doi.org/10.1055/s-0039-1677897.
16. Escourrou, P., Luriau, S., Rehel, M., Nédelcoux, H. and Lanoë, J. L. (2000) 'Needs and costs of sleep monitoring', *Studies in health technology and informatics*, 78, p. 69–85. Available at: http://europepmc.org/abstract/MED/11151608.
17. Abdel-Basset, M., Ding, W. and Abdel-Fatah, L. (2020) 'The fusion of Internet of Intelligent Things (IoIT) in remote diagnosis of obstructive Sleep Apnea: A survey and a new model', *Information Fusion*. Elsevier B.V., 61(April), pp. 84–100. https://doi.org/10.1016/j.inffus.2020.03.010.
18. Bianchi, M. T. (2018) 'Sleep devices: wearables and nearables, informational and interventional, consumer and clinical', *Metabolism: Clinical and Experimental*. Elsevier Inc., 84, pp. 99–108. https://doi.org/10.1016/j.metabol.2017.10.008.
19. Choi, Y. K., Demiris, G., Lin, S. Y., Iribarren, S. J., Landis, C. A., Thompson, H. J., McCurry, S. M., Heitkemper, M. M. and Ward, T. M. (2018) 'Smartphone applications to support sleep self-management: Review and evaluation', *Journal of Clinical Sleep Medicine*, 14(10), pp. 1783–1790. https://doi.org/10.5664/jcsm.7396.
20. Lin, C. T., Prasad, M., Chung, C. H., Puthal, D., El-Sayed, H., Sankar, S., Wang, Y. K., Singh, J. and Sangaiah, A. K. (2017) 'IoT-Based Wireless Polysomnography Intelligent System for Sleep Monitoring', *IEEE Access*. IEEE, 6, pp. 405–414. https://doi.org/10.1109/ACCESS.2017.2765702.
21. Azimi, I., Oti, O., Labbaf, S., Niela-Vilen, H., Axelin, A., Dutt, N., Liljeberg, P. and Rahmani, A. M. (2019) 'Personalized maternal sleep quality assessment: An objective iot-based longitudinal study', *IEEE Access*, 7, pp. 93433–93447. https://doi.org/10.1109/ACCESS.2019.2927781.
22. Saleem, K., Bajwa, I. S., Sarwar, N., Anwar, W. and Ashraf, A. (2020) 'IoT Healthcare: Design of Smart and Cost-Effective Sleep Quality Monitoring System', *Journal of Sensors*, 2020. https://doi.org/10.1155/2020/8882378.
23. Mendonça, F., Mostafa, S. S., Morgado-Dias, F. and Ravelo-García, A. G. (2020) 'An oximetry based wireless device for sleep apnea detection', *Sensors (Switzerland)*, 20(3). https://doi.org/10.3390/s20030888.
24. Sadek, I. and Mohktari, M. (2018) 'Nonintrusive Remote Monitoring of Sleep in Home-Based Situation', *Journal of Medical Systems*. Journal of Medical Systems, 42(4), pp. 1–10. https://doi.org/10.1007/s10916-018-0917-6.
25. Wu, Y., Zhou, J., Li, J., Liu, J., Li, S. and Bai, C. (2018) 'Application of IoT-based medical diagnosis and treatment in patients with obstructive sleep apnea/hypopnea syndrome in primary hospitals: A preliminary study', *Traditional Medicine and Modern Medicine*, 01(03), pp. 207–212. https://doi.org/10.1142/s257590001850012x.
26. Yacchirema, D., Sarabia-Jácome, D., Palau, C. E. and Esteve, M. (2018b) 'System for monitoring and supporting the treatment of sleep apnea using IoT and big data', *Pervasive and Mobile Computing*, 50, pp. 25–40. https://doi.org/10.1016/j.pmcj.2018.07.007.

27. Tal, A., Shinar, Z., Shaki, D., Codish, S. and Goldbart, A. (2017) 'Validation of contact-free sleep monitoring device with comparison to polysomnography', *Journal of Clinical Sleep Medicine*, 13(3), pp. 517–522. https://doi.org/10.5664/jcsm.6514.
28. Liao, L. De, Wang, Y., Tsao, Y. C., Wang, I. J., Jhang, D. F., Chuang, C. C. and Chen, S. F. (2020) 'Design and implementation of a multifunction wearable device to monitor sleep physiological signals', *Micromachines*, 11(7). https://doi.org/10.3390/mi11070672.
29. Burgdorf, A., Güthe, I., Jovanović, M., Kutafina, E., Kohlschein, C., Bitsch, J. Á. and Jonas, S. M. (2018) 'The mobile sleep lab app: An open-source framework for mobile sleep assessment based on consumer-grade wearable devices', *Computers in Biology and Medicine*, 103, pp. 8–16. https://doi.org/10.1016/j.compbiomed.2018.09.025.
30. Baty, F., Boesch, M., Widmer, S., Annaheim, S., Fontana, P., Camenzind, M., Rossi, R. M., Schoch, O. D. and Brutsche, M. H. (2020) 'Classification of sleep apnea severity by electrocardiogram monitoring using a novel wearable device', *Sensors (Switzerland)*, 20(1), 1–12. https://doi.org/10.3390/s20010286.
31. Milici, S., Lazaro, A., Villarino, R., Girbau, D. and Magnarosa, M. (2018) 'Wireless Wearable Magnetometer-Based Sensor for Sleep Quality Monitoring', *IEEE Sensors Journal*, 18(5), pp. 2145–2152. https://doi.org/10.1109/JSEN.2018.2791400.
32. Zhang, N., Li, Y., Xiang, S., Guo, W., Zhang, H., Tao, C., Yang, S. and Fan, X. (2020) 'Imperceptible sleep monitoring bedding for remote sleep healthcare and early disease diagnosis', *Nano Energy*. Elsevier Ltd, 72(February), p. 104664. https://doi.org/10.1016/j.nanoen.2020.104664.
33. Behar, J., Roebuck, A., Shahid, M., Daly, J., Hallack, A., Palmius, N., Stradling, J. and Clifford, G. D. (2015) 'SleepAp: An automated obstructive sleep apnoea screening application for smartphones', *IEEE Journal of Biomedical and Health Informatics*. IEEE, 19(1), pp. 325–331. https://doi.org/10.1109/JBHI.2014.2307913.
34. Petrov, M. E., Hasanaj, K., Hoffmann, C. M., Epstein, D. R., Krahn, L., Park, J. G., Hollingshead, K., Yu, T. Y., Todd, M., St. Louis, E. K., Morgenthaler, T. I. and Buman, M. P. (2020) 'Rationale, design, and development of SleepWell24: A smartphone application to promote adherence to positive airway pressure therapy among patients with obstructive sleep apnea', *Contemporary Clinical Trials*, 89(November 2019). https://doi.org/10.1016/j.cct.2019.105908.
35. Castillo-Escario, Y., Ferrer-Lluis, I., Montserrat, J. M. and Jane, R. (2019) 'Entropy analysis of acoustic signals recorded with a smartphone for detecting apneas and hypopneas: A comparison with a commercial system for home sleep apnea diagnosis', *IEEE Access*, 7, pp. 128224–128241. https://doi.org/10.1109/ACCESS.2019.2939749.
36. Werner-Seidler, A., O'Dea, B., Shand, F., Johnston, L., Frayne, A., Fogarty, A. S. and Christensen, H. (2017) 'A Smartphone App for Adolescents With Sleep Disturbance: Development of the Sleep Ninja', *JMIR Mental Health*, 4(3), p. e28. https://doi.org/10.2196/mental.7614.
37. Borger, J. N., Huber, R. and Ghosh, A. (2019) 'Capturing sleep–wake cycles by using day-to-day smartphone touchscreen interactions', *npj Digital Medicine*. Springer US, 2(1), pp. 1–8. https://doi.org/10.1038/s41746-019-0147-4.
38. Ren, Y., Wang, C., Chen, Y., Yang, J. and Li, H. (2019) 'Noninvasive Fine-Grained Sleep Monitoring Leveraging Smartphones', *IEEE Internet of Things Journal*. IEEE, 6(5), pp. 8248–8261. https://doi.org/10.1109/JIOT.2019.2922283.
39. Cuttone, A., Bækgaard, P., Sekara, V., Jonsson, H., Larsen, J. E. and Lehmann, S. (2017) 'SensibleSleep: A Bayesian model for learning sleep patterns from smartphone events', *PLoS ONE*, 12(1). https://doi.org/10.1371/journal.pone.0169901.
40. Scott, H., Lack, L. and Lovato, N. (2018) 'A pilot study of a novel smartphone application for the estimation of sleep onset', *Journal of Sleep Research*, 27(1), pp. 90–97. https://doi.org/10.1111/jsr.12575.

41. Shin, H. and Cho, J. (2014) 'Unconstrained snoring detection using a smartphone during ordinary sleep', *BioMedical Engineering Online*, 13(1), 1–14. https://doi.org/10.1186/1475-925X-13-116.
42. Arora, S., Baig, F., Lo, C., Barber, T. R., Lawton, M. A., Zhan, A., Rolinski, M., Ruffmann, C., Klein, J. C., Rumbold, J., Louvel, A., Zaiwalla, Z., Lennox, G., Quinnell, T., Dennis, G., Wade-Martins, R., Ben-Shlomo, Y., Little, M. A. and Hu, M. T. (2018) 'Smartphone motor testing to distinguish idiopathic REM sleep behavior disorder, controls, and PD', *Neurology*, 91(16), pp. E1528–E1538. https://doi.org/10.1212/WNL.0000000000006366.
43. Gu, W., Shangguan, L., Yang, Z. and Liu, Y. (2016) 'Sleep hunter: Towards fine grained sleep stage tracking with smartphones', *IEEE Transactions on Mobile Computing*. IEEE, 15(6), pp. 1514–1527. https://doi.org/10.1109/TMC.2015.2462812.
44. Mehmood, Y., Ahmad, F., Yaqoob, I., Adnane, A., Imran, M. and Guizani, S. (2017) 'Internet-of-Things-Based Smart Cities: Recent Advances and Challenges', *IEEE Communications Magazine*. IEEE, 55(9), pp. 16–24. https://doi.org/10.1109/MCOM.2017.1600514.
45. Robinson Joel, M., Ebenezer, V., Karthik, N. and Rajkumar, K. (2019) 'Advance dynamic network system of internet of things', *International Journal of Recent Technology and Engineering*, 8(3), pp. 6209–6212. https://doi.org/10.35940/ijrte.C5657.098319.
46. Gubbi, J., Buyya, R., Marusic, S. and Palaniswami, M. (2013) 'Internet of Things (IoT): A vision, architectural elements, and future directions', *Future Generation Computer Systems*. Elsevier B.V., 29(7), pp. 1645–1660. https://doi.org/10.1016/j.future.2013.01.010.
47. Pinheiro, E., Postolache, O. and Girão, P. (2010) 'Theory and Developments in an Unobtrusive Cardiovascular System Representation: Ballistocardiography', *The Open Biomedical Engineering Journal*, 4(1), pp. 201–216. https://doi.org/10.2174/1874120701004010201.
48. Jararweh, Y., Tawalbeh, L., Ababneh, F. and Dosari, F. (2013) 'Resource efficient mobile computing using cloudlet infrastructure', *Proceedings - IEEE 9th International Conference on Mobile Ad-Hoc and Sensor Networks, MSN 2013*, pp. 373–377. https://doi.org/10.1109/MSN.2013.75.
49. Wan, J., Zou, C., Ullah, S., Lai, C. F., Zhou, M. and Wang, X. (2013) 'Cloud-Enabled wireless body area networks for pervasive healthcare', *IEEE Network*. IEEE, 27(5), pp. 56–61. https://doi.org/10.1109/MNET.2013.6616116.
50. Kim, J.-Y., Chu, C.-H. and Kang, M.-S. (2020) 'IoT based Unobtrusive Sensing for Sleep Quality Monitoring and Assessment', *IEEE Sensors Journal*, 21(3), pp. 1–1. https://doi.org/10.1109/jsen.2020.3022915.
51. Chang, X., Tang, X., Chen, L., Liu, A., Wang, Yan, Liu, N. and Wang, Yurong (2021) 'Hospital IoT big data and real-time monitoring and nursing intervention for patients with insomnia', *Microprocessors and Microsystems*. Elsevier B.V., 81(December 2020), p. 103667. https://doi.org/10.1016/j.micpro.2020.103667.
52. Ma, B., Wu, Z., Li, S., Benton, R., Li, D., Huang, Y., Kasukurthi, M. V., Lin, J., Borchert, G. M., Tan, S., Li, G., Yang, M. and Huang, J. (2020) 'Development of a support vector machine learning and smart phone Internet of Things-based architecture for real-time sleep apnea diagnosis', *BMC Medical Informatics and Decision Making*. BioMed Central, 20(14), pp. 1–13. https://doi.org/10.1186/s12911-020-01329-1.
53. Yacchirema, D. C., Sarabia-Jacome, D., Palau, C. E. and Esteve, M. (2018a) 'A smart system for sleep monitoring by integrating IoT with big data analytics', *IEEE Access*, 6, pp. 35988–36001. https://doi.org/10.1109/ACCESS.2018.2849822.
54. Peake, J. M., Kerr, G. and Sullivan, J. P. (2018) 'A critical review of consumer wearables, mobile applications, and equipment for providing biofeedback, monitoring stress, and sleep in physically active populations', *Frontiers in Physiology*, 9(JUN), pp. 1–19. https://doi.org/10.3389/fphys.2018.00743.
55. Fitbit Inc. (2017) *Fitbit Official Site for Activity Trackers and More, fitbit.com*. Available at: https://www.fitbit.com/in/home (Accessed: 29 January 2021).
56. Oura (2020) *Oura Ring: Accurate Health Information Accessible to Everyone, ouraring.com*. Available at: https://ouraring.com/ (Accessed: 29 January 2021).

57. Dreem (2021) *Dreem 2 - Sleep, finally*. Available at: https://dreem.com/ (Accessed: 29 January 2021).
58. Advanced Brain Monitoring (2014) *Sleep profiler*. Available at: https://www.advancedbrain-monitoring.com/products/sleep-profiler#section-overview (Accessed: 29 January 2021).
59. Compumedics Limited (2021) *Somté – Compumedics*. Available at: https://www.compumed-ics.com.au/en/products/somte/ (Accessed: 29 January 2021).
60. *General Sleep Corporation* (n.d.). Available at: https://www.generalsleep.com/zmachine-synergy.html (Accessed: 29 January 2021).
61. Corporation, G. S. (2021) *General Sleep Corporation*. Available at: https://www.gener-alsleep.com/zmachine-insight.html (Accessed: 29 January 2021).
62. Sleep Shepherd (2021) *Sleep Improvement - Sleep Shepherd*. Available at: https://sleepshep-herd.com/sleep-improvement/ (Accessed: 29 January 2021).
63. Re-Timer (2020) *Home - Re-Timer*. Available at: https://www.re-timer.com/ (Accessed: 29 January 2021).
64. AYO (2021) *Meet AYO: Award Winning Blue Light Emitting Wearable Technology*. Available at: https://goayo.com/ayo-glasses (Accessed: 29 January 2021).
65. Bose (2020) *Hush Support | Bose*. Available at: https://www.bose.com/en_us/support/hush_support.html (Accessed: 29 January 2021).
66. Kokoon (2021) *How it works | Kokoon*. Available at: https://kokoon.io/pages/how-it-works (Accessed: 29 January 2021).
67. DreamPad (2021) *Why it Works - Dreampad*. Available at: https://dreampadsleep.com/pages/why-it-works (Accessed: 29 January 2021).
68. Apple Inc (2019) *Beddit Sleep Monitor*. Available at: https://www.beddit.com/ (Accessed: 29 January 2021).
69. CPAP (2021) *CPAP machines - Find the best CPAP machine for you - ResMed*. Available at: https://www.resmed.com/en-us/sleep-apnea/cpap-products/cpap-machines/ (Accessed: 29 January 2021).
70. Hexoskin (2018) *Hexoskin Smart Shirts - Cardiac, Respiratory, Sleep & Activity Metrics, Hexoskin*. Available at: https://www.hexoskin.com/ (Accessed: 29 January 2021).
71. Bian, J., Guo, Y., Xie, M., Parish, A. E., Wardlaw, I., Brown, R., Modave, F., Zheng, D. and Perry, T. T. (2017) 'Exploring the association between self-reported asthma impact and fitbit-derived sleep quality and physical activity measures in adolescents', *JMIR mHealth and uHealth*, 5(7). https://doi.org/10.2196/mhealth.7346.
72. Haghayegh, S., Khoshnevis, S., Smolensky, M. H., Diller, K. R. and Castriotta, R. J. (2020) 'Performance assessment of new-generation Fitbit technology in deriving sleep parameters and stages', *Chronobiology International*, 37(1), pp. 47–59. https://doi.org/10.1080/07420528.2019.1682006.
73. Moreno-Pino, F., Porras-Segovia, A., López-Esteban, P., Artés, A. and Baca-García, E. (2019) 'Validation of fitbit charge 2 and fitbit alta hr against polysomnography for assessing sleep in adults with obstructive sleep apnea', *Journal of Clinical Sleep Medicine*, 15(11), pp. 1645–1653. https://doi.org/10.5664/jcsm.8032.
74. Ramagopalan, S. V., McDonald, L. and Mehmud, F. (2018) 'Sleep and BMI: Do (Fitbit) bands aid?', *F1000Research*, 7. https://doi.org/10.12688/f1000research.14774.2.
75. Chaudhry, F. F., Danieletto, M., Golden, E., Scelza, J., Botwin, G., Shervey, M., De Freitas, J. K., Paranjpe, I., Nadkarni, G. N., Miotto, R., Glowe, P., Stock, G., Percha, B., Zimmerman, N., Dudley, J. T. and Glicksberg, B. S. (2020) 'Sleep in the natural environment: A pilot study', *Sensors (Switzerland)*, 20(5). https://doi.org/10.3390/s20051378.
76. Koskimäki, H., Kinnunen, H., Kurppa, T. and Röning, J. (2018) 'How do we sleep - A case study of sleep duration and quality using data from oura ring', in *UbiComp/ISWC 2018 - Adjunct Proceedings of the 2018 ACM International Joint Conference on Pervasive and Ubiquitous Computing and Proceedings of the 2018 ACM International Symposium on Wearable Computers*, pp. 714–717. https://doi.org/10.1145/3267305.3267697.

77. de Zambotti, M., Baker, F., Rosas, L., Claudatos, S. and Colrain, I. (2017) '0111 A POLYSOMNOGRAPHIC VALIDATION STUDY OF A NOVEL COMMERCIALLY-AVALIABLE MULTISENSORY SLEEP TRACKER', *Sleep*, 40(suppl_1), pp. A41–A42. https://doi.org/10.1093/sleepj/zsx050.110.
78. Arnal, P. J., Thorey, V., Ballard, M. E., Hernandez, A. B., Guillot, A., Jourde, H., Harris, M., Guillard, M., van Beers, P., Chennaoui, M. and Sauvet, F. (2019) 'The dreem headband as an alternative to polysomnography for EEG signal acquisition and sleep staging', *bioRxiv*. https://doi.org/10.1101/662734.
79. Arnal, P. J., Thorey, V., Debellemaniere, E., Ballard, M. E., Hernandez, A. B., Guillot, A., Jourde, H., Harris, M., Guillard, M., van Beers, P., Chennaoui, M. and Sauvet, F. (2020) 'The dreem headband compared to polysomnography for electroencephalographic signal acquisition and sleep staging', *Sleep*, 43(11). https://doi.org/10.1093/sleep/zsaa097.
80. Guillot, A., Sauvet, F., During, E. H. and Thorey, V. (2020) 'Dreem Open Datasets: Multi-Scored Sleep Datasets to Compare Human and Automated Sleep Staging', *IEEE Transactions on Neural Systems and Rehabilitation Engineering*, 28(9). https://doi.org/10.1109/TNSRE.2020.3011181.
81. Puchkova, A. N., Tkachenko, O. N., Trapeznikov, I. P., Piletskaya, I. A., Tiunova, E. V., Sazonova, M. M., Taranov, A. O., Gruzdeva, S. S. and Dorokhov, V. B. (2019) 'Assessment of potential capabilities of Dreem: An ambulatory device for EEG phase-locked acoustic stimulation during sleep', *SOCIALNO-ECOLOGICHESKIE TECHNOLOGII*, 9(1). https://doi.org/10.31862/2500-2961-2019-9-1-96-112.
82. Saganowski, S., Kazienko, P., Dziezyc, M., Jakimów, P., Komoszynska, J., Michalska, W., Dutkowiak, A., Polak, A., Dziadek, A. and Ujma, M. (2020) 'Review of consumer wearables in emotion, stress, meditation, sleep, and activity detection and analysis', *arXiv*.
83. Finan, P. H., Richards, J. M., Gamaldo, C. E., Han, D., Leoutsakos, J. M., Salas, R., Irwin, M. R. and Smith, M. T. (2016) 'Validation of a Wireless, Self-Application, Ambulatory Electroencephalographic Sleep Monitoring Device in Healthy Volunteers', *Journal of Clinical Sleep Medicine*, 12(11), pp. 1443–1451. https://doi.org/10.5664/jcsm.6262.
84. Levendowski, D. J., Gamaldo, C., St Louis, E. K., Ferini-Strambi, L., Hamilton, J. M., Salat, D., Westbrook, P. R. and Berka, C. (2019) 'Head position during sleep: Potential implications for patients with neurodegenerative disease', *Journal of Alzheimer's Disease*, 67(2), pp. 631–638. https://doi.org/10.3233/JAD-180697.
85. Scott, H., Lack, L. and Lovato, N. (2020) 'A systematic review of the accuracy of sleep wearable devices for estimating sleep onset', *Sleep Medicine Reviews*, p. 6. https://doi.org/10.1016/j.smrv.2019.101227.
86. Ellender, C., Swieca, J. and Cunnington, D. (2015) 'Accuracy of non-invasive sleep monitors compared with polysomnography', *Sleep and Biological Rhythms*, 13.
87. Wolfhope, K. and Hudkins, A. (2016) 'Increasing Quality of Sleep Utilizing the iLs Dreampad Mini in Children With Autism Spectrum Disorder', *American Journal of Occupational Therapy*, 70(4_Supplement_1), p. 7011515252p1. https://doi.org/10.5014/ajot.2016.70s1-po2023.
88. Gutman, S. A., Gregory, K. A., Sadlier-Brown, M. M., Schlissel, M. A., Schubert, A. M., Westover, L. A. and Miller, R. C. (2017) 'Comparative effectiveness of three occupational therapy sleep interventions: A randomized controlled study', *OTJR Occupation, Participation and Health*, 37(1), pp. 5–13. https://doi.org/10.1177/1539449216673045.
89. Schoen, S. A., Man, S. and Spiro, C. (2017) 'A Sleep Intervention for Children with Autism Spectrum Disorder: A Pilot Study', *The Open Journal of Occupational Therapy*, 5(2). https://doi.org/10.15453/2168-6408.1293.
90. Srikolcheep, N. and Sittiprapaporn, P. (2017) 'Efficacy of the Integrated Listening Systems' Dreampad™ device to Sleep Quality in insomnia patient', in *2nd Joint International Conference on Digital Arts, Media and Technology 2017: Digital Economy for Sustainable Growth, ICDAMT 2017*, pp. 356–359. https://doi.org/10.1109/ICDAMT.2017.7904991.

91. Jones, C. and Moyle, W. (2020) 'A feasibility study of Dreampad™ on sleep, wandering and agitated behaviors in people living with dementia', *Geriatric Nursing*, 41(6), pp. 782–789. https://doi.org/10.1016/j.gerinurse.2020.04.014.
92. Lain, D. and Bourn, S. (2009) 'Quality assurance report on the use of continuous positive airway pressure and end-tidal carbon dioxide during respiratory distress in field emergency care', *Critical Care*, 13(Suppl 1). https://doi.org/10.1186/cc7170.
93. Mishra, A., Singh, M., Kamboj, S., Prabhu, S. and Stapleton, D. (2014) 'Evaluation of Monitoring of Continuous Positive Airway Pressure (CPAP) Therapy Compliance in Primary Care Setting: A Quality Initiative Project', *Chest*, 146(4). https://doi.org/10.1378/chest.1984889.
94. Damien Leger, Maxime Elbaz, Benoît Piednoir, Amélie Carron, Joëlle Texereau (2016) 'Evaluation of the add-on NOWAPI® medical device for remote monitoring of compliance to Continuous Positive Airway Pressure and treatment efficacy in obstructive sleep apnea', *BioMedical Engineering Online*, 15(1).
95. PSFK (2018) 'How Brands Like Spire Health Are Developing Wearable Workout Optimization Tools', *PSFK*.
96. Slamon, N., Nadkarni, V., Penfil, S. and Parker, R. (2018) '241: ANALYSIS OF HEART RATE AND HEART RATE VARIABILITY DURING LIVE PEDIATRIC INTENSIVE CARE ACTIVITIES', *Critical Care Medicine*, 46(1). https://doi.org/10.1097/01.ccm.0000528260.44703.62.
97. Ferrer-Lluis, I., Castillo-Escario, Y., Montserrat, J. M. and Jane, R. (2020) 'Analysis of Smartphone Triaxial Accelerometry for Monitoring Sleep-Disordered Breathing and Sleep Position at Home', *IEEE Access*, 8, pp. 71231–71244. https://doi.org/10.1109/ACCESS.2020.2987488.
98. Lee, P. H., Marek, J. and Nálevka, P. (2020) 'Crowdsourced smartphone data reveal altered sleep/wake pattern in quarantined Chinese during the COVID-19 outbreak', *Chronobiology International*. Taylor & Francis, 37(8), pp. 1181–1190. https://doi.org/10.1080/07420528.2020.1792486.
99. Philip, P., Dupuy, L., Morin, C. M., de Sevin, E., Bioulac, S., Taillard, J., Serre, F., Auriacombe, M. and Micoulaud-Franchi, J. A. (2020) 'Smartphone-based virtual agents to help individuals with sleep concerns during COVID-19 confinement: Feasibility study', *Journal of Medical Internet Research*, 22(12), pp. 1–14. https://doi.org/10.2196/24268.
100. Tiron, R., Lyon, G., Kilroy, H., Osman, A., Kelly, N., O'Mahony, N., Lopes, C., Coffey, S., McMahon, S., Wren, M., Conway, K., Fox, N., Costello, J., Shouldice, R., Lederer, K., Fietze, I. and Penzel, T. (2020) 'Screening for obstructive sleep apnea with novel hybrid acoustic smartphone app technology', *Journal of Thoracic Disease*, 12(8), pp. 4476–4495. https://doi.org/10.21037/jtd-20-804.
101. Sanjiv Narayan, Priyanka Shivdare, Tharun Niranjan, Kathryn Williams, Jon Freudman, and R. Sehra (2019) 'HHS Public Access', *Physiology & behavior*, 176(3), pp. 139–148. https://doi.org/10.1007/s11325-018-1695-6. Noncontact.
102. Patel, P., Kim, J. Y. and Brooks, L. J. (2017) 'Accuracy of a smartphone application in estimating sleep in children', *Sleep and Breathing*, 21(2), pp. 505–511. https://doi.org/10.1007/s11325-016-1425-x.
103. Bhat, S., Ferraris, A., Gupta, D., Mozafarian, M., De Bari, V. A., Gushway-Henry, N., Gowda, S. P., Polos, P. G., Rubinstein, M., Seidu, H. and Chokroverty, S. (2015) 'Is there a clinical role for smartphone sleep apps? Comparison of sleep cycle detection by a smartphone application to polysomnography', *Journal of Clinical Sleep Medicine*, 11(7), pp. 709–715. https://doi.org/10.5664/jcsm.4840.
104. Nakano, H., Hirayama, K., Sadamitsu, Y., Toshimitsu, A., Fujita, H., Shin, S. and Tanigawa, T. (2014) 'Monitoring sound to quantify snoring and sleep apnea severity using a smartphone: Proof of concept', *Journal of Clinical Sleep Medicine*, 10(1), pp. 73–78. https://doi.org/10.5664/jcsm.3364.

105. Ahad, A., Tahir, M., Sheikh, M. A., Ahmed, K. I., Mughees, A. and Numani, A. (2020) 'Technologies trend towards 5g network for smart health-care using iot: A review', *Sensors (Switzerland)*, 20(14), pp. 1–22. https://doi.org/10.3390/s20144047.
106. Yu, H., Lee, H. and Jeon, H. (2017) 'What is 5G? Emerging 5G mobile services and network requirements', *Sustainability (Switzerland)*, 9(10). https://doi.org/10.3390/su9101848.
107. Al-Falahy, N. and Alani, O. Y. (2017) 'Technologies for 5G Networks: Challenges and Opportunities', *IT Professional*, pp. 12–20. https://doi.org/10.1109/MITP.2017.9.
108. *Benefits of 5G: Advantages and features | About Verizon* (n.d.) *Verizon*. Available at: https://www.verizon.com/about/news/benefits-of-5g (Accessed: 3 February 2021).
109. Carvalho, G., Cabral, B., Pereira, V. and Bernardino, J. (2021) 'Edge computing: current trends, research challenges and future directions', *Computing*. https://doi.org/10.1007/s00607-020-00896-5.
110. Stephanie Allen, P., Hammett, D. R., de Vettori, E., Purdy, L., Qun, Z., Krolop, S., Sehgal, C., Wada, Y., Bergen, M. Van, Suboh, A., Loke, D. W. C., Siegel, S. and Burrill, S. (2019) '2019 Global health care outlook Shaping the future', *Design Issues*, p. 41. Available at: https://www2.deloitte.com/content/dam/Deloitte/global/Documents/Life-Sciences-Health-Care/gx-lshc-hc-outlook-2019.pdf%0A; http://www.ncbi.nlm.nih.gov/books/NBK2665/%0A; https://doi.org/10.1016/j.bios.2016.09.038%0A; https://pdfs.semanticscholar.org/956b/6ee61.
111. Liu, X., Jia, M., Zhang, X. and Lu, W. (2019) 'A Novel Multichannel Internet of Things Based on Dynamic Spectrum Sharing in 5G Communication', *IEEE Internet of Things Journal*, 6(4), pp. 5962–5970. https://doi.org/10.1109/JIOT.2018.2847731.
112. Vega, M. (2018) 2020's Internet of Things Statistics, Facts & Predictions. Available at: https://review42.com/internet-of-things-stats/ (Accessed: 4 February 2021).
113. *Global Wearable Computing Devices Market (2020 to 2025)—Growth, Trends & Forecasts* (2020) *Business Insider*. Available at: https://www.globenewswire.com/news-release/2020/06/24/2052588/0/en/Global-Wearable-Computing-Devices-Market-2020-to-2025-Growth-Trends-Forecasts.html (Accessed: 4 February 2021).
114. O'Dea, S (2020) • *5G mobile subscriptions forecast worldwide 2019-2025 | Statista, statista*. Available at: https://www.statista.com/statistics/521598/5g-mobile-subscriptions-worldwide/ (Accessed: 4 February 2021).
115. Hang, L. and Kim, D.-H. (2019) 'Design and implementation of an integrated iot blockchain platform for sensing data integrity', *Sensors*. Multidisciplinary Digital Publishing Institute, 19(10), p. 2228.
116. Khosla, S., Deak, M. C., Gault, D., Goldstein, C. A., Hwang, D., Kwon, Y., O'Hearn, D., Schutte-Rodin, S., Yurcheshen, M., Rosen, I. M. and others (2018) 'Consumer sleep technology: an American Academy of Sleep Medicine position statement', *Journal of clinical sleep medicine*. American Academy of Sleep Medicine, 14(5), pp. 877–880.
117. Behar, J. A., Liu, C., Kotzen, K., Tsutsui, K., Corino, V. D. A., Singh, J., Pimentel, M. A. F., Warrick, P., Zaunseder, S., Andreotti, F. and others (2020) 'Remote health diagnosis and monitoring in the time of COVID-19', *Physiological measurement*. IOP Publishing, 41(10), p. 10TR01.
118. Ahmad, I., Kumar, T., Liyanage, M., Okwuibe, J., Ylianttila, M. and Gurtov, A. (2018) 'Overview of 5G security challenges and solutions', *IEEE Communications Standards Magazine*. IEEE, 2(1), pp. 36–43.

Chapter 12
Case Studies on 5G and IoT Security Issues from the Leading 5G and IoT System Integration Vendors

S. C. Sandeep, Thenmozhi Rayan, Kumudavalli, and Sathish Kumar

12.1 Introduction

The Internet of Things (IoT) is completely reshaping the life of an individual and society with its strong influence on every section of society, and even the lifestyle of tech-savvy has been dramatically modified. Its influence can be seen in realizing simple applications like home automation to government initiatives of smart city projects. The Indian government has initiated dream smart city projects to modify the existing cities to smart cities incorporating digitalization and IoT ideas. IoT is a network of smart edge computing nodes exchanging information via the internet between the local edge devices and the Cloud. The underlying 5G infrastructure can support millions of devices per square kilometer compared to nearly hundreds of thousands of devices with preceding technologies. There is a tremendous increase in the number of computing nodes connected and controlled through IoT technology in recent years. Each node in the network has its own defined application, but yet they exhibit certain functional similarities.

IoT establishes an integrated ecosystem of the surrounding environment and network-based systems resulting in improved performance, financial benefits, and optimized human effort. The key driving force behind IoT's development is MOSFETs, which formed a fundamental building block leading to miniaturization

S. C. Sandeep (✉) · T. Rayan
Dayananda Sagar College of Engineering, Bangalore, India

Kumudavalli
Dayananda Sagar College of Arts, Science and Commerce, Bangalore, India
e-mail: kumudavalli@dayanandasagar.edu

S. Kumar
Department of Electrical Engineering and Computer Science, Cleveland State University, Cleveland, OH, USA
e-mail: s.kumar13@csuohio.edu

S. Velliangiri et al. (eds.), *Secure Communication for 5G and IoT Networks*, EAI/Springer Innovations in Communication and Computing,
https://doi.org/10.1007/978-3-030-79766-9_12

of the electronic device and increased transistor density in the integrated chips. Many applications could be embedded on smart devices that are very small in size, less in weight, and consumes very little power. IoT definition has evolved due to integrating several technologies, such as real-time data analytics, extensive data analysis, cognitive systems, machine learning, artificial intelligence, etc.

The problem called "basket of remotes" was predicted by the Apple alumni team and BeOS committees co-founder. Many applications must interact with several devices functioning with different protocols. The first autonomous device network idea was conceived as early as 1982 by Coca-Cola Company for its vending machine design at Carnegie Mellon University. The communication between edge computing nodes was conceptualized in 1999 by Bill Joy while presenting his Six Webs framework at World Economic Forum at Davos. The IoT technology gained momentum and entered most of the social and commercial domains introducing a paradigm shift in the technology world.

As per the estimation of Cisco Systems, the growth rate of the things to people ratio has increased from 0.08 in 2003 to 1.84 in 2010. The year-on-year growth of 30% is estimated in the number of IoT devices. The 7.8 trillion markets are estimated by 2020. There will be about 50 billion devices connected to the internet by 2020, according to the prediction of Cisco and Ericcson [1]. IoT would encode about 50–100 trillion objects and also able to track their movement. Every human in an urban area is surrounded by about 1000–5000 trackable objects. The sudden surge in the implementation of IoT in various fields poses serious challenges in privacy and security. Due to lack of business planning deployment of about 70–75% of getting stuck in the prototype or pilot stage, according to a 2018 study. The devices' location can be tracked, and information is communicated accordingly to provide effective utilization of devices. However, the communication of a large amount of data to handle the application due to variable spatial scale constraints is a challenging factor. Therefore, the geographical diversity in standards plays a crucial role in IoT devices' internet and web applications. The growth of IoT is further driven by the evolution of 5G technology, which provides extended coverage, lower latency, and higher throughput and high-density connection to network devices (Fig. 12.1).

IoT implementation to improve efficiency depends on the ease of access to store and process data. To achieve this, companies collect and acquire data from multiple sources through IoT and store the information on their cloud network for further processing and future usage. Hence, personal data becomes vulnerable to misuse and poses privacy and security issues during transmission and storing. Also, the user loses control and ownership of personal data and its usage in the future. Therefore, in the issues like a massive amount of data generated, establishing large-scale networks, and heterogeneity management [2], QoS requirement is very challenging in real-time. To address this issue, governments are forming regulations related problems. However, these may change from one country to another, which further complicates the implementation and development of applications on the World Wide Web. The IoT environment's complexity is due to a considerable number of

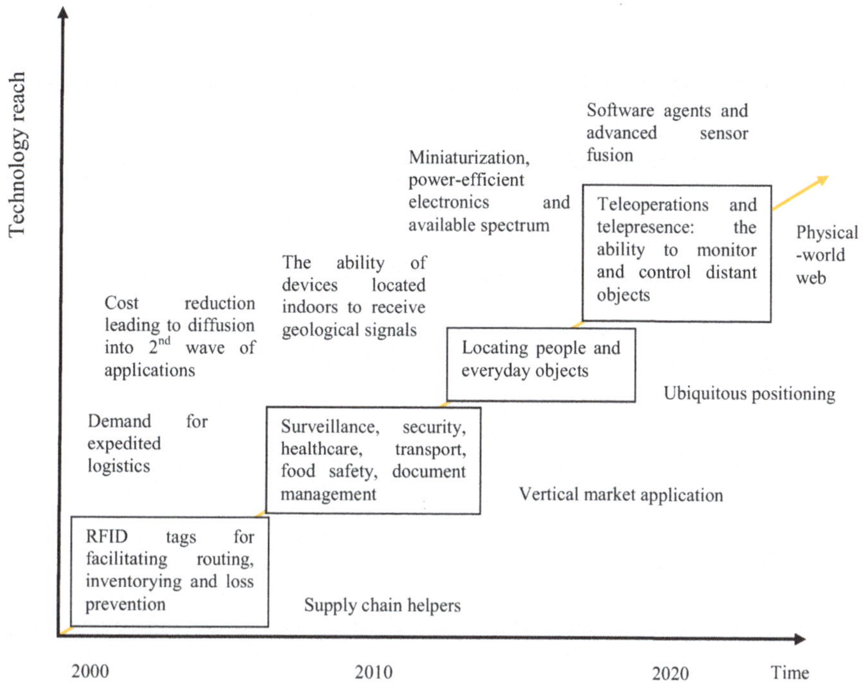

Fig. 12.1 Technology roadmap: the Internet of Things

different links operating simultaneously and the interaction of autonomous devices with the Cloud.

Federal Trade Commission (FTC) published a report in January 2015 consisting of three recommendations.

- Data security—During the design cycle, the companies dealing in IoT-related services should ensure the security of the data collected, stored, and processed. Also, the security features and policies are to be upgraded for better security. At the same time, they also should adopt the government policies described by the local/regional government. Companies should adopt a "defense-in-depth" approach and encrypt data at each stage.
- Data consent—The companies should offer the user flexibility to choose the data they wish to share with service companies. Also, while collecting or sharing the collected data with any third party should implement user consent procedures. The provision should also be made to modify and delete the data by the user depending on the services opted by them.
- Data minimization—The service-providing companies should collect the minimum amount of data required to provide an opted service. The user must be notified about the time for which the collected data validity is in force. The minimization of data collection improves the users' trust and convenience and optimizes the network bandwidth usage. Especially, in regions with weak networks,

connection management could be an issue to collect extensive data at once. The user may be prompted to provide the same data repeatedly, leading to the user's frustration. In some cases, the user may also opt out of the service.

The following risk factors and opportunities were examined by the World Bank in government adoption of IoT and published a report including the following points.

- The government may still take some more time before opting to implement IoT-based infrastructure and services.
- The regulatory policies are to be developed entirely before adopting.
- The business model does not have a clear picture about various parameters such as security and integrity of the data although there is a strong value proposition to implement.
- The organizational and capacity gap between the government and the private sector companies to be addressed in consent.
- Consistency in data validation and management is missing with many service providers.
- Governments and companies need to implement infrastructure to implement IoT effectively and efficiently.
- Governments take control of enabling and disabling certain services and data collection. Any government may not be keen on the citizens' share data with multinational or foreign companies.
- Most pilots have typical characteristics of public–private partnership or partnering with local technological leaderships.

12.1.1 Intelligence

Companies like Intel researched the concept of the integration of IoT with autonomous control systems. Deep reinforcement learning provides a dynamic and interactive environment to process elements; thus, process elements cannot implement the supervised algorithm as they use conventional machine learning methodology. The intelligence to IoT environment can be offered in three levels: IoT/network devices, Edge computing nodes, and cloud computing. The sensitivity of the application and data availability demands the need for intelligent data processing and control flow at each level. The researchers are trying to integrate advanced machine learning algorithms like deep learning into IoT devices to make smart objects a reality. IoT edge devices or things have limited processing and storage capacity and are distributed over the entire network. These limits lead to integration with cloud computing (CC) [2], which provides enormous storage and high processing capability. These advantages are achieved based on shared resources to maximize the overall operational ability of the Cloud unit. The future smart devices connected to IoT networks may adapt and work on an open network environment in which self-organized autonomous entities and virtual elements may be interoperable depending on context, circumstances, and environments. Many efforts are being made to mitigate the measurement errors to improve the accuracy in decision-making based on the collection and reasoning of the information's context. The smart device, which is automated to choose different technologies and solutions based on the context of operation, is already available. The smart devices are classified into ambient data collections and analytical devices like blood alcohol content (BAC) and real-time reactive applications like autonomous vehicles [3]. The technological constraints and limitations can be compensated by virtually unlimited capabilities of data processing and analytical tools and resources of Cloud, hence providing seamless operational access to edge devices over IoT network.

5G I-IoT (Intelligent IoT) offers intelligent data processing and optimized communication channel usage with the convergence of intelligence, the internet, and devices. The higher processing capabilities and large bandwidth availability in the 5G infrastructure enable edge devices to communicate a large amount of data to the cloud core for processing and accurate decision-making.

12.1.2 Architecture

A simple three-layer view can represent IoT architecture. Each layer operates autonomously yet sharing and exchanging information with other layers for better control and decision-making. The network is built with colossal scalability to handle a surge of devices. The service-oriented architecture (SOA) can be used in IoT implementation constituted with four layers of development with distinct features [4].

Layer 1—Devices: Includes networked objects such as sensors, actuators, and other devices communicating over public or proprietary protocols with edge gateways. Every IoT system is based on the data captured and provided by edge devices. These devices actually capture the real-time data from the world and convert it into digital data for further processing and analysis. These can be embedded into a more extensive system, or they can be independent standalone devices.

Layer 2—Edge processing elements (EPEs) or gateways: Provides abstract interfacing between layer one devices and layer three clouds for easy, seamless operations and establishes secured connection using systems like Web Socket. In some applications, this layer is also designed to provide preprocessing of data. These elements must have a bidirectional communication facility to operate efficiently and accurately. The implementation offers a more significant challenge when these elements are resource-constrained and battery-operated.

Layer 3—Cloud: Includes the applications built for IoT using microservice architecture and various database systems to store the data received from different EPEs with high-security features. The Cloud services need to be general and compatible with different topologies. The services provided by Cloud are essential as they are the ones providing the services and data processing.

Layer 4—Interface: Provides methods to interface users and applications. This layer integrates the services and applications provided by the companies to a user.

The applications and services may work on SaaS architecture, and respective details are captured in the Cloud.

The Cloud architecture can be different depending on the context, and also, the number of logical layers may vary [5]. Aneka helps developers build Microsoft .Net applications with the support of APIs. Architectural advantages are used as a service by the developers. The upper layers of Cloud generally include "Infrastructure as a Service (IaaS), Platform as a Service (PaaS), and Software as a Service (SaaS)."

The interface design needs to be user-friendly. At the same time, they should be best integrated into the IoT environment. The widespread recognition and usage of IoT demand evolving design technologies and scalability to support future demands.

The entire architecture of 5G I-IoT implementation can be viewed in three categories [6] as:

- Processing center: Usually, a cloud-based core has a multi-access facility to interact with many devices spread over the region.
- Object processor: The local processing unit installed in the edge device or base station to process the raw data capture and transmit to the processing unit for further analysis.
- Sensing region: This is a physical region covered by the sensor network having a device to devise connectivity (D2D).

12.1.3 Addressability

The devices were initially identified using Auto-ID based on RFID tags and Electronics Product Codes. Over the period, advanced addressing methods like IP address or "URI (Uniform Resource Identifier)" is used to identify devices. The IPv4 provides an address space for up to 4.3 billion different addresses. The next-generation IPv6 has scaled to a vast address space required for IoT. This protocol also provides auto-configuration, enabling the devices over the network to configure themselves without much interference from the hosts.

As the number of devices over the network increases exponentially, addressing them would be challenging in the future. Soon the ability of IPv4 is exhausted, and the entire infrastructure may have to migrate to IPv6. Then, the device address has to be updated for billions of devices to maintain the compatibility with Cloud. Each device address should be revised remotely as physically reaching each one of them to update is not the practically feasible solution, nor it is economical. In some cases, the devices would have integrated into larger systems and not accessible at all.

The current IPv6 assignment policy involves existing LIR (Local Internet Registry) with IPv4 allocation from APNIC (Asia Pacific Network Information Centre). This provisioning will make IPv6 assignment possible. Addressing using IPv6 also provides a technical advantage along with a more significant number of devices. It provides a hierarchical address allocation method that limits the size of routing tables. The design of IPv6 is adhered closely with existing IPv4 and provided end-to-end datagram transmission.

12.1.4 Standards and Standards Organizations

Mozilla launched project Things in 2017 to create a decentralized IoT by giving URLs on the web to make devices link and discover. A standard data model and APIs are defined to make them interoperable. Web of things is a unifying application layer for IoT linking multiple underlying IoT protocols to existing web technologies. Mozilla is a member of the Web of Things Interest Group currently undergoing standardization at the W3C (World Wide Web Consortium). The standard organizations aspiring to set up IoT infrastructure using open standards are listed with respective technical standards.

Organization name	Standard under development
Auto Identification Center	Networked Radiofrequency identification (RFID)
Electronic Product Code Technology	Electronic Product Code (EPC) technology adaptation standards
US Food and Drug Administration	For medical devices, the "Unique Device Identification (UDI)" system is used
Global Standards One	The standard for fast-moving consumer goods (FMCG) using UDI and RFID
Institute of Electrical and Electronics Engineers	Underlying communication technology standards such as IEEE 802.15.4, IEEE P1451-99 (IoT Harmonization), and IEEE P1931.1 (ROOF Computing)
Internet Engineering Task Force	Standards comprising TCP/IP
MTConnect Institute	Data exchange standards for medicine tools and related industrial equipment
Open Data Format	Open Data format is a standard that specifies generic information on model structure declared by the Internet of Things (IoT) Work Group of The Open Group in 2014
Open Messaging Interface	The limited set of critical operations required for IoT systems are specified and provided by O-MI is a standard as published by the Internet of Things Work Group of The Open Group in 2014
Open Connectivity Foundation	Constrained Application Protocol (CoAP) standard for simple devices
Open Mobile Alliance	"Open Mobile Alliance Device Management (OMA-DM)" and Open Mobile Alliance Light Weight M2M (OMALWM2M) for IoT device management and GotAPI for a secure framework
XMPP Standards Foundation	For instant messaging, "Extensible Messaging and Presence Protocol" is defined which is an open standard protocol
Third Generation Partnership Projects (3GPP)	Develop global communication standards classifies use cases into Enhanced mobile broadband (eMBB), Ultra-reliable low-latency communications (uRLLC), Massive machine-type communication (mMTC)

12.1.5 IoT Core and Platform Fragmentation

The services like Cloud ML, Data Studio, and DataLab are integrated by Google IoT core using big data and machine learning analysis. The Azure IoT Hub, which is used for core services by Google and Amazon, offers services in two tiers. The primary option has limited services offering only unidirectional communication capabilities from device to Cloud. The standard option activates and allows all the available features for IoT solutions for bidirectional communication. The free-tier version of Azure has a limitation on the device count accessing the Cloud per unit time and on the message count per day. The maximum number of devices that can connect to the Cloud is 100 per second. The ratio of broker service time to the number of messages published per second is similar in Amazon and Google cloud services. Direct subscriptions are not allowed in Google and Microsoft platforms. Instead, they forward all messages to additional services. Microsoft has developed an open IoT platform for experimental research and development activities at academic institutions called Lab of Things [2]. It uses HomeOS4 to perform activities such as interconnecting various devices, implementing application scenarios, deploying case studies, monitoring field implications, and sharing data. The Windows-based PC Home Hub application assists the users to communicate with the public cloud platform.

The platform fragmentation results from variation in the underlying hardware components on which the system is built. Each device may be operating on different hardware, and the execution capability of each one of them varies over a wide range. A similar problem arises in software. There are many operating systems incorporating various algorithms for the efficient and effective operation of the system. Hence, the platform fragmentation is a big challenge in implementing IoT infrastructure. The design of applications working consistently between inconsistent technologies ecosystems becomes exceedingly difficult and complex in such scenarios. The amorphous computing nature also poses security problems. According to one research, 87% of Android active devices tend to fail. They cannot update to the latest version in the absence of new software patches. The compatibility for the devices to adopt the latest services must be provided as patches. Also, revised security features have to be communicated timely to the edge devices to secure the operations. The Messaging and registration of shadow devices are billed based on regions in the Amazon Web Service IoT core. Prices in Google core are billed based on the amount of data used for a service period every month and offer free services for enabling, accessing, editing, and deleting devices.

Ericsson provides 5G core (5GC) as a combination of Evolved Packet Core (EPC), and 5GC network functions on a common multi-access cloud platform which provides efficient Total Cost of Ownership (TCO) and a smooth migration to 5G.

12.1.6 Data Storage

The challenges in data management are the vast amount of data being generated and satisfying the data acquisition requirements of different environment applications. The NoSQL database solution offers better performance in applications where the number of data to be stored is dynamic as in the case of IoT [7]. The number of sensors transmitting the data to the Cloud changes over time. Therefore, the amount of data to be stored in Cloud also changes dynamically.

The IoT offers two fundamental challenges to service providers.

- To perform operations like removal, analysis, and understand the large amount of data captured by the sensors nodes or edge devices.
- To provide storage of this bulk data with optimal power consumption, the internet is already consuming about 5% of total energy generated.

Chronos is a library containing around 90 inbuilt libraries for third-party customer services that enhance user experience on specific services [2]. This library consists of services such as "Yahoo weather, eBay product shopping, Facebook Graph API, Flickr photo" management, "Twitter microblogging, Twilio telephony, Amazon cloud, PayPal payment, Uber vehicle confirmation, YouTube video streaming, Google analytics" to name a few.

Oracle IoT provides a combination of Open-that performs acquisition. Insight-that analysis the data, Secure-that relates the safety, and Accelerate-that integrates data [2]. These four parameters are a crucial part of an Oracle IoT cloud.

IBM IoT cloud is a well-developed layered architecture that provides security and ease device connectivity to the platform [2]. It provides connectivity to independent chips to intelligent appliances and advanced applications. "Identity as a Service (IDaaS)" is the backbone of IBM IoT cloud, which hosts and manages Cloud as a "Single Sign-On (SSO)" service. The lightweight MQTT messaging protocol can be used to attach "AMR embed, Arduino, Intel Galileo etc., to IBM cloud." Developers can create applications with RESTful and real-time APIs to connect data coming from devices to IBM Bluemix. IBM Bluemix also performs Hadoop governed IBM analytics and geospatial analytics. In general device management-related operations, such as upgrade, diagnostics, inspection, and monitoring of device, firmware are offered by IBM Cloud.

The main components of Google IoT Core are the device manager which includes registration, authentication and authorization processes, and the protocol bridge [8]. The device manager's registration of devices depends on the required necessary services opted by the user. These devices communicate with the Cloud using protocol bridges such as HTTP and MQTT. Devices use MQTT to send requests to a specific topic, and HTTP maintains the platform's connection. The Google platform is comparatively more straightforward for the user to define the information as required.

12.1.7 *Security and Services*

Security is a significant concern in adopting IoT without appropriate consideration of issues associated. However, the technical security concerns remain the same as in conventional standalone applications, such as weak authentication, retaining default passwords or not changing them frequently, unencrypted message exchanges between the devices, and poor handling of available security features updates. The computational operations available in devices are limited because they cannot implement a firewall or use strong cryptosystems. Also, the developers are constrained to design devices with lower cost and power consumption which makes implementation of robust security features difficult. Poorly secured devices are subverted to attacks by spyware and malware to extract personal and private data from IoT devices. IoT security foundation (IoTSF) was launched in 2015 founded by technology providers and telecommunication companies to promote associated knowledge and best practices. Innovative services are researched and developed by information technology companies to provide the best security solutions possible.

As per KBV research, due to the diversification in IoT and increase in infrastructural concerns, the IoT security services market is estimated to grow at a CAGR of 27.9% from 2016 to 2022. The comprehensive suite is provided by Etherios, which supports products and services for the connected enterprise. PaaS model is used to design its Cloud to enable user access to any product in real-time. By combining machine data with associated Salesforce.com instances, Social Machine can provide SaaS-based solutions and transform itself into a powerful CR004 [2]. The device authentication under public/private mode for higher-order security using JSON Web Tokens is provided by Google IoT Core [8]. RSA or Elliptic Curve algorithms are used to verify data signatures. Root certificate authorities are communicated using TLS 1.2 protocol for communication security. The control, authentication, and authorization of Cloud IoT API access are managed by Google Cloud "Identity and Access Management (IAM)."

The introduction of the IMSI encryption method, 5G, marked a new era of network security, under which the data integrity is protected with mutual authentication between device and network. The implementation and realization of functionalities on 5G are not standardized by the authorities such as 3GPP but is agnostic that it is flexible enough to allow an overlap of different physical and virtual radio access network (RAN) and core network. This leads to the adoption of end-to-end encryption as an integral tool to ensure system security and disciplined system design principles.

12.1.8 Safety

Sensor inputs, user data, or other external stimuli are usually the inputs to the intelligent apps on IoT systems. Based on these inputs, respective commands are executed to provide different types of automation. There are few specific IoT problems, such as buggy applications, bad application interaction, or device and communication failures that can cause unsafe and dangerous physical states. Identifying such defects demands a holistic approach of interaction between installed applications, physical devices, and their configurations. A novel practical system called IotSan is recently proposed by the "University of California Riverside" that the model checking is a building block to disclose "interaction-level" flaws by identifying events that can drive the system to dangerous states.

12.1.9 IoT Adoption Barriers

There are some significant obstacles in adopting IoT across the technological ecosystem due to the complexity of providing unified service. Each electronic device has been designed for a specific job, and its operation may ultimately differ from other devices around. Also, there are many different communication protocols used across the user platform. Hence, it is vital to establish a common platform to communicate over IoT infrastructure. The IoT network consists of numerous nodes through which collected user information is passed. The diversity of the devices in the network poses serious concerns about the data's security and privacy. Fundamentally, IoT architecture consists of four security objectives.

- Data confidentiality: The transmitted and stored data are inaccessible to unauthorized parties or devices.
- Data integrity: The intentional corruption of data has to be detected, and appropriate action has to be initiated against the body initiating the corruption. Also, there can be an unintentional corruption of transmitted data which can be retransmitted and stored for future usage.
- Non-repudiation: The sender is held responsible for the data provided and cannot deny it under any circumstances.
- Data availability: Authorized parties are given access to transmitted and stored data even if there is denial of service (DOS) attacks.

12.1.10 The Role of IoT Edge Computing

IoT edge devices solve the problem associated with the centralization of cloud architecture. Although clouds operate faster as they are much powerful for storage and processing, many delays are introduced when devices are continuously sent and received. By introducing IoT edge devices, the architecture enables distributed computing which provides many advantages such as faster data process, prevent delays, security breaches, and other network-related concerns [9].

In many real-time applications, the delay caused due to sending data to the Cloud and processing may lead to catastrophic effects. Local processing helps in real-time business systems and alerts and manages the data sent to the Cloud.

Core IoT edge architecture's five components are listed subsequently.

- Complex and fast event processing unit.
- Cloud implementation of machine learning (ML) and artificial intelligence (AI) models for understanding the user requirements over the period.
- Applications to support user requirements and services.
- Online and off-line support for better service.
- Data management, including security and integrity issues.

Edge devices also support various networking options such as Ethernet, wireless networks like Wi-Fi, and next-generation wireless network 5G. The aesthetic design of these devices supports harsh environments due to their hardened and durable casing.

12.1.11 The Major Components of IoT Edge Computing Are

The mobile devices and edge computing architecture provide an alternative path to cloud networking for data transport.

- Complex event processing (CEP): The operational technology deployments are using CEP software and services for decades. These systems can acquire raw data from various sensors and process the raw data according to the specific pattern. Apache Storm is one of the standard open-source technology for building and running CEP systems.
- Machine learning: This feature enables machines to understand the significance and importance of data and build required models using known algorithms. However, the machines can be made to learn patterns and anomalies by their own using advanced machine learning tools.
- Artificial intelligence: Google search is an example of how AI is used to sophisticate the machine learning process. Based on words, the search engine understands the meaning and maps to the related images and videos. During the process, the machine learns the meaning of a word typed by the user. Integrating AI technology in IoT devices enables advanced services like identifying a person visiting the premises.
- Applications: IoT devices having powerful hardware processing capabilities can run containerized applications, including CEP, AI, ML, and other tools. These applications become a more critical element when data is being processed locally, and decisions are made based on results.

- Off-line data: The IoT devices include sufficient storage capabilities to suffer network connection issues in the deployed area. The data stored can then be transmitted to the Cloud once the network is restored.
- Data management: The validation of captured data is essential to improve the efficiency of the device. This has to be done locally to avoid unnecessary transmissions over the network. Also, it helps in reducing network congestions, especially in localities where bandwidth is constrained.

12.2 Present and Future IoT Devices

The flexibility to adapt to upgraded applications and enabling remote configuration of edge devices are vital aspects of future devices. The automatic deployment of updates provides better security and improves device capabilities. The specialized applications and services required to improve the capabilities of the device are developed by the vendors. The Cloud available currently does not adhere to the predefined standard format of data representation or process handling. Standardization in future can help in removing the obstructions for service providers and live customers by improving interoperability. Currently, all IoT devices cannot connect or operate on a shared Cloud due to their design's heterogeneity, which introduces network delays and data loss. An abstraction needs to be developed to address the issue to enhance the network's performance and scalability. The content awareness computing technique is to be developed to process data collected by billions of different devices over the network. Developing the middleware to enable the horizontal flow of data, unlike clouds following vertical silos, can incorporate multi-localized modules. The system is presently using IPv4 numbered addresses that will get exhausted in few years to come, and devices should incorporate a new addressing policy such as IPv6. Energy harvesting is another important aspect which many devices fail to address in current situations. The data flow is wholly interrupted due to power failure in one device in the entire data flow chain from the Edge device to the Cloud. Depleted batteries may also lead to faulty operations of sensors and devices. Suppose the devices are made capable of harvesting renewable energies such as solar or wind energy. In that case, this issue can be addressed.

Artificial intelligence (AI) is expected to play a significant role in future IoT implementation to perform intelligent tasks such as voice recognition, language translation, decision-making, and so on. AI can significantly help to accrue the considerable amount of data processed by IoT devices about users. The synthesis of two concepts can make three fundamental analyses based on data collected to provide customized services.

1. Predictive—This analysis helps in preemptive industrial application intervention to reduce machinery failures and improve operational safety for users.
2. Prescriptive—Based on the scenarios, the appropriate course of action can be provided to improve disaster management capabilities.

3. Adaptive—Continuous data collected by the sensors feed can help the system take frequent actions autonomously without human intervention.

The convergence of blockchain, AI, and IoT can effectively drive the future wave of digital transformation and open new business models. Blockchain can increase the trust, transparency, security, and privacy of the business by providing a shared and distributed management model. Years ago, blockchain was discussed only in financial transactions. However, recently, more and more non-financial use cases of blockchain technology have emerged due to the collective implementation of all three concepts.

The 5G network will be necessary for future IoT implementations as the tremendous amount of data flow demands proportionately high network speed and bandwidth. 5G spectrum will expand the frequencies on which digital data are transferred. Ultra-high bandwidth of 5G can enhance the use of Augmented and Virtual Reality (AR/VR).

Software-Defined Network (SDN) completely shifts the paradigm of network management by decoupling data and control planes. This provides flexibility and programmability in the IoT network without affecting the functional architecture of existing implementation.

12.3 Comparative Analysis

The IoT services are provided by various companies at different levels of infrastructure, but there are specific essential parameters to be considered which define the core of IoT services to analyze the impact of those services on the end-user. End-user acceptance is of prime importance for revenue generation. In turn, it is also the driving force for further innovations. The following comparison table provides comprehensive information about IoT services offered by companies.

Companies	Services offered at various layers of IoT infrastructure
Google	1. The platform for end-user devices supports the seamless operation of a wide variety of embedded OSs with Debian Linux OS 2. The devices manufactured by leading companies like Intel and Microchip are supported with real-time out-of-box services 3. Google's server-less infrastructure running on Cloud IoT Core has the real-time scaling capability with stringent industry-standard security protocols 4. TLS1.2 based on Asymmetric key authentication for end-to-end security 5. Industry-standard protocols such as MQTT and HTTP are supported for Protocol Bridge 6. Big data analytical and ML services such as "Cloud Dataflow, BigQuery, Cloud Bigtable, ML, Google Data Studio"

Companies	Services offered at various layers of IoT infrastructure
IBM	1. IBM offers a Watson IoT platform to manage cloud-hosted services, which derives value from IoT devices. It also includes additional services like Blockchain service and analytic service 2. IBM Maximo: Provides essential insights for intelligent asset management and operations 3. IBM TRIRIGA: Efficient solutions for real estate and facilities management 4. IBM Maximo Application Suite uses Enterprise Asset Management (EAM) to optimize performance using AI and analytics 5. Inventory-related cost is reduced and optimized by maintenance, repair, and operation (MRO) solution
Microsoft	1. Azure IoT central accelerates the creation of IoT services 2. IoT solutions offered by Azure enable remote monitoring, predictive maintenance, and device simulation 3. IoT Edge is a complete managed service built at the Azure hub 4. Azure Digital Twins creates comprehensive models for physical environments 5. Azure time series insights improve operation and decision-making with massive IoT data delivered 6. Azure Sphere is a comprehensive IoT security service that includes hardware, OS, and cloud components 7. Azure maps deliver geospatial APIs to include maps, analytics, and mobility solutions 8. Azure RTOS implements high-performance devices with reliability on embedded systems
Cisco	1. Cisco Kinetic IoT platform maximizes value for extracted data 2. Cisco DNA Centre and Field Network Director are used to integrate IT and OT infrastructure 3. IoT routers are optimized with the deployment of Cisco IR11001 integrated services router rugged 4. Cisco also extends cybersecurity services and data Intelligence to the IoT edge devices/nodes 5. Cisco cyber vision integrates It and OT services to ensure productivity, resilience, and safety
Amazon	1. AWS provides end-to-end services from edge devices to the Cloud and brings data management and analytics together 2. Offers services at all layers of security and preventive measures like encryption and data control 3. AWS IoT integrates with other services of AWS to build a complete IoT solutions package 4. AWS IoT Greengrass software enables local computing, data caching, sync, and machine learning capabilities 5. AWS Device defender continuously monitors and audits IoT configurations to ensure best security practices 6. Device registration is managed remotely using the AWS IoT device management tool

12.4 Conclusion

The study intends to highlight the characteristics of different platforms and services to have a broader understanding. The current implementation of IoT infrastructure needs to address the heterogeneity management issue at the core. The management of device, data, and applications seems to favor and encourage upscaling of various nodes. The three most popular cloud platforms study that none is perfect in every scenario for developers—some lack open-source APIs and some security aspects. Data management-based platforms capture the maximum market, while deployment management captures the most negligible market value. The remaining aspects, such as device and application management, heterogeneity, analysis of the statistical data, observing various operations, anticipation, and research domain, are between ranges.

References

1. Carsten Maple "Security and privacy in the internet of things", Journal of Cyber Policy, 2:2, 155–184.
2. Partha Pratim Ray "A survey of IoT cloud platforms", Future Computing and Informatics Journal 1 2016, pp. 35–46.
3. Anne H. Ngu, Mario Gutierrez, Vangelis Metsis, Surya Nepal, and Quan Z. Sheng "IoT Middleware: A Survey on Issues and Enabling Technologies", IEEE Internet Of Things Journal, Vol. X, No. X, March 2016.
4. Li Da Xu, Wu He, Shancang Li "Internet of Things in Industries: A Survey", IEEE Transactions on Industrial Informatics, Volume: 10, Issue: 4, Nov. 2014.
5. Karthikeyyan, P., & Velliangiri, S. (2019, July). Review of blockchain-based IoT application and its security issues. In 2019 2nd International Conference on Intelligent Computing, Instrumentation and Control Technologies (ICICICT) (Vol. 1, pp. 6–11). IEEE.
6. Dan Wang, Dong Chen, Bin Song, Nadra Guizani, Xiaoyan Yu, and Xiaojiang Du, "From IoT to 5G I-IoT: The Next Generation IoT-Based Intelligent Algorithms and 5G Technologies" IEEE Communications Magazine • October 2018
7. L. Incipini, A. Belli, L. Palma, R. Concetti, and P. Pierleoni, "Databases performance evaluation for IoT systems: The Scrovegni chapel use case" in Proc. 42nd Int. Conv. Inf. Commun. Technol., Electron. Microelectron. (MIPRO), May 2019, pp. 463–468.
8. Paola Pierleoni, Roberto Concetti, Alberto Belli, and Lorenzo Palma, "Amazon, Google and Microsoft Solutions for IoT: Architectures and a Performance Comparison", IEEE open Access, December 23, 2019.
9. V. Anbarasu, P. Karthikeyan and S. P. Anandaraj, "Turning Human and Food Waste into Reusable Energy in a Multilevel Apartment Using IoT," 2020 6th International Conference on Advanced Computing and Communication Systems (ICACCS), Coimbatore, India, 2020, pp. 440–444, https://doi.org/10.1109/ICACCS48705.2020.9074170.

Chapter 13
A Saving Energy MANET Routing Protocol in 5G

Khanh Quy Vu, Vijender Kumar Solanki, and Anh Ngoc Le

13.1 Introduction

MANETs are a set of mobile radio devices capable of self-configuration, self-established parameters to transmission data in-network without relying on preexisting infrastructure [1, 2]. Although limited in ability and capabilities, MANETs have demonstrated superiority in communication with flexible infrastructures, promising a significant contribution to the next-generation networks in the future [3, 4].

Along with the evolution of mobile communication systems, the fifth generation (5G) mobile network is beginning to be commercially deployed. The advent of 5G marks the beginning of a truly digital society with breakthrough improvements in data transfer speed, mobility, and energy efficiency [5]. Moreover, according to Cisco expected, by 2023, over 13.1 billion mobile devices will be linked to the Internet. In particular, these devices are installed IoT modules (the principle of formation to MANETs in 5G) [6]. These devices consume a massive amount of electrical energy. Therefore, the issue of energy saving in MANETs in 5G is particularly interesting [2]. It requires the improvement of existing systems [7, 8]. In the MANETs environment, when a network device is exhausted, it affects the performance and the network lifetime of the overall system. Consequently, many works have been researching to propose saving energy routing protocols for MANETs [9, 10]. Survey results in [5–10] show that the energy efficiency research direction is

K. Q. Vu
Hung Yen University of Technology and Education, Hungyen, Vietnam
e-mail: quyvk@utehy.edu.vn

V. K. Solanki
CMR Institute of Technology, Hyderabad, Telangana, India

A. N. Le (✉)
Electric Power University, Hanoi, Vietnam
e-mail: anhngoc@epu.edu.vn

S. Velliangiri et al. (eds.), *Secure Communication for 5G and IoT Networks*,
EAI/Springer Innovations in Communication and Computing,
https://doi.org/10.1007/978-3-030-79766-9_13

divided into two main approaches: (1) Power Control and (2) Network Lifetime. Survey results also showed that the network lifetime-based approach has been interested in research and achieved positive results. However, there is still no solution or class of algorithms that can be applied to all MANET scenarios. Each protocol has advantages and limitations and only suitable for an individual scenario. However, existing solutions can be combined to provide a more energy efficiency mechanism.

In this paper, we proposed an energy-based routing protocol for the MANETs in 5G, use the combined metric of hops number and remaining energy of the node to select the suitable path, The proposed protocol is improved from DSR aim enhance the performance and the network lifetime. In the next section, we demonstrate our proposed protocol. Simulation results are analyzed and presented in Sect. 13.3, and Sect. 13.4 is the Conclusion.

13.2 Proposed Routing Protocol

Survey results show that routing protocols proposed for MANETs must reflect the throughput, delay, reliability, and QoS metrics. In this work, we focus on determining the parameters that directly affect the network lifetime, thereby proposing an energy-saving routing protocol, called EED (energy-efficient DSR) routing protocol. According to [2], the energy consumption of a wireless node dependent on two main activities:

- Sending or receiving data packets.
- Listen from neighboring nodes on the wireless environment.

We propose taking two metrics into account in making decision is the hops number and the remaining battery capacity of the node. Specific calculations are presented in the next sections.

13.2.1 Protocol Description

Like routing protocols DSR [11] and AODV [12], the EED protocol includes procedures to discovery and maintain routes, as follows:

The route discovery procedure: whenever a source node needs transmission packets, it will send broadcast RREQ packets with header modified {MinEnergy, TotalEnergy, DSR RREQ Header) to find a route to the destination network node. Then, RREQs will reach the destination node through intermediate nodes. At each intermediate node, when received the RREQs, these nodes will perform Energy – check procedure, in Fig. 13.1. This procedure has two main tasks:

- Determine the lowest remaining energy level in the route.
- Determine the total remaining energy in the route.

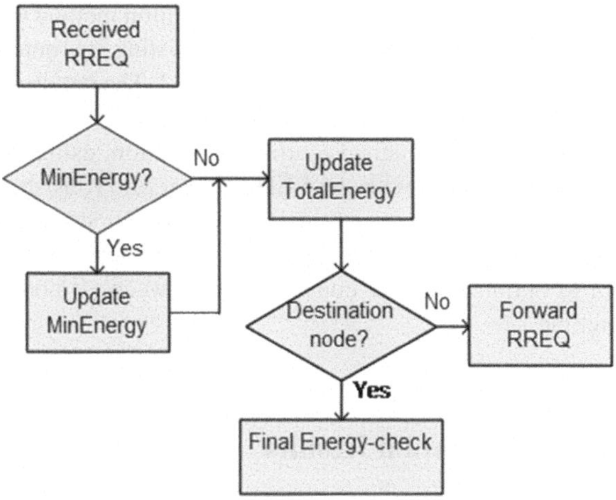

Fig. 13.1 Energy-check procedure.

Finally, the destination node will determine the optimal route based on routing mtrics in RREQs and sends the RREP packet to the source network node. Besides, this protocol also has the error route update procedure using RRERs (Route Error) packets. In this way, the source node received the fit route [13].

13.2.2 Routing Method

Then the destination node is the received candidate routes set. This work defines two constraints, as follows:

- The hops number of routes must be within [Hopmin, Hopmax]. Routes that have hops number out of range will be discarded.

$$\text{Hopnumber} = [\text{Hopmin}, \text{Hopmax}] \tag{13.1}$$

where Hopmin is the hops number of the shortest route and Hopmax is defined by:

$$\text{Hopmax} = \text{Hopmin} + 2 \tag{13.2}$$

- Aim to determine the route that has the richest energy; this work defines the cost function as follows:

$$\text{EED}_{(P)} = \min_{1 \le j \le P} \left(\frac{E_{\max}}{E_t + 1} \right) + \sum_{P}^{i=1} \left(\frac{E_{\max} / E_t}{P} \right) \tag{13.3}$$

An illustration of the EED cost calculation method with different values is shown in Table 13.1. Assuming that there are existing six routes, each node has the remaining battery capacity, as shown in Fig. 13.2. The battery capacity of *S* and *D* nodes is equal to 5/10, Hopmin = 2 and Hopmax = 4.

Based on the obtained routing information, using Eq. (13.3), the EED protocol will select route 6 with the EED value = 2.111, as shown in Table 13.1. With our approach, a route between source-destination node pairs is always determined based on a balance between two factors: hops number and energy. It helps the EED protocol to optimize energy consumption between all nodes and improve the network lifetime.

13.3 Simulation Results

In this subsection, we establish simulations to evaluate the system performance based on the criteria: *average remaining energy of node*, *network lifetime*, and *packet delivery ratio* of three protocols are DSR, AODV, and EED on the NS-2 simulation software. Our simulation system includes 200 mobile nodes, randomly arranged in a square of 2.000 m × 2.000 m. Simulations that use traffic type are CBR; IEEE 802.11 standard; packet size are 512 (Byte), and initial energy of nodes is installed at 8 (J) according to the two simulation scenarios, as follows:

Scenario 1: Simulations are done with 25 CBR connections for a period of 350 (s) to consider the average remaining energy.

Scenario 2: CBR connections are: 5, 10, 15, 20, 25, 30, 35, and 40, respectively. The simulation parameters are summarized in Table 13.2.

Figure 13.3 presents the system performance based on the metric: *Average Remaining Energy*. Results showed that the energy consumption for the first 100 (s) of all three protocols is relatively stable and equivalent. This consumption tends to increase faster towards the end of the simulation. However, the average remaining Energy of Node of the EED protocol is always higher than the DSR and AODV protocols in the overall simulation process. Results show that thanks to the combined load distribution mechanism and find the shortest path, the EED protocol improves energy consumption compared to the traditional protocols.

Table 13.1 Method of calculating the route cost

Routes	Hops	Average energy	Min energy	EED
1	3	1.667	2.000	3.667
2	4	1.458	2.000	3.458
3	4	1.333	1.667	3.000
4	4	1.190	1.250	2.440
5	5	1.146	1.250	2.396
6	**5**	**1.000**	**1.111**	**2.111**

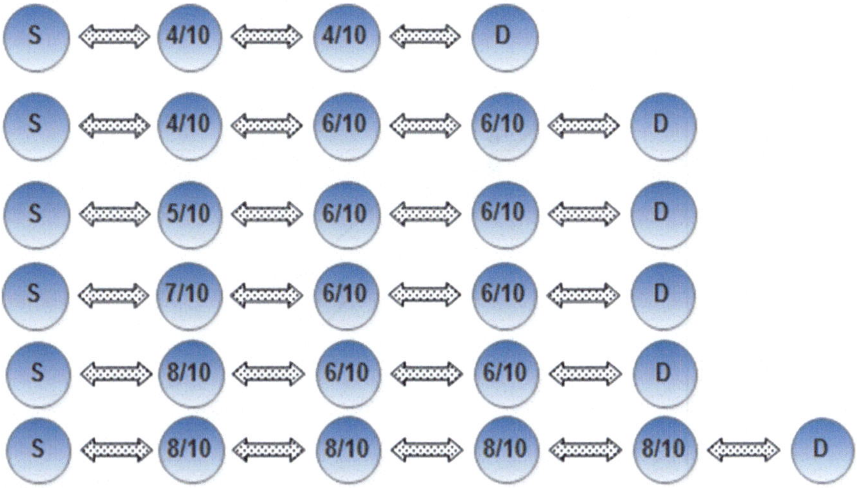

Fig. 13.2 An Illustrated candidate routes after the discovery procedure

Table 13.2 Simulation parameters

Parameter	Value
Number nodes	200
Mobile speed	1 (m/s)
Transmission power	1.0 (W)
Receive power	0.1 (W)
Overhearing power	0.02 (W)
Transmission range	250 (m)
Hopmax	Hopmin + 2
Mobility model	Random Waypoint
Simulation time	350 (s)

Figure 13.4 presents the system performance based on the metric: Network Lifetime. Results showed that EED significantly improved compared to the traditional protocols. These achieved results thank the ability of the load distribution on entire nodes of the EED algorithm which led to preventing overloaded nodes or abused.

Figure 13.5 presents the system performance based on the metric: *Packet Delivery Ratio*. Results showed that the packet delivery ratio tends to decrease when connection numbers increase. When connection numbers are less than 10, the packet delivery ratio of protocols is equivalent. However, when the network traffic increases, the proposed protocol significantly improves the packet delivery ratio compared to traditional protocols. Observing the results show that the EED protocol improves most of the performance metrics when the connection number is over ten pairs.

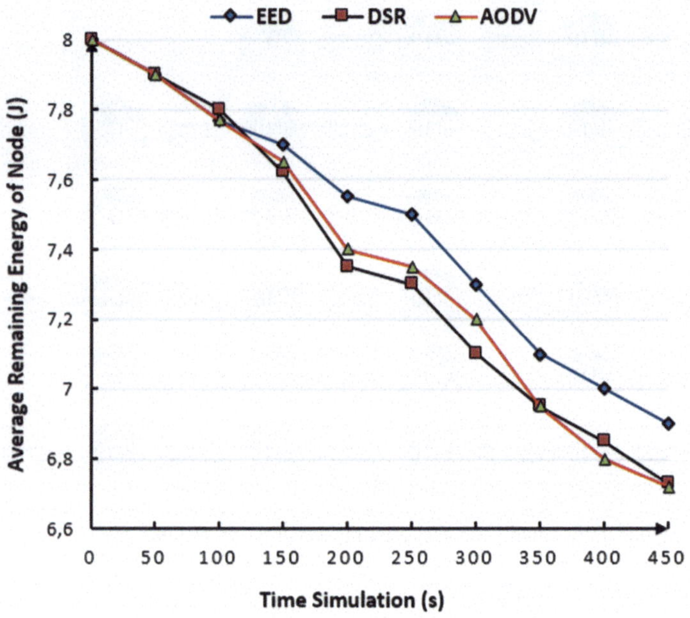

Fig. 13.3 Remaining energy in Scenario 1

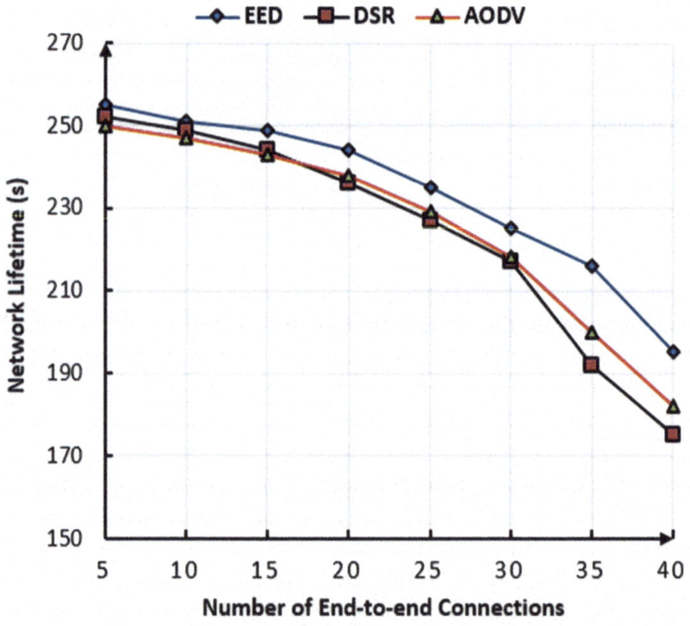

Fig. 13.4 Network lifetime in Scenario 2

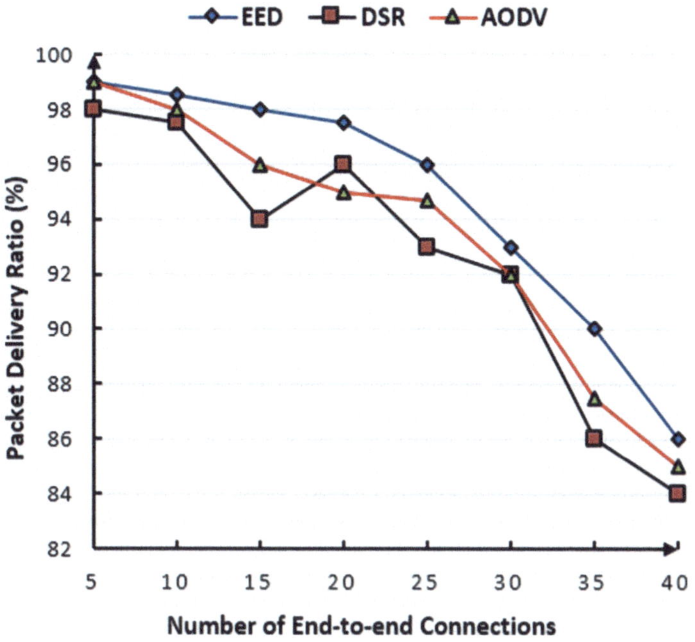

Fig. 13.5 Packet delivery ratio in Scenario 2

13.4 Conclusions

In this work, we proposed a new energy-based routing protocol for the MANET in 5G, called EED, using a load distribution routing mechanism with a cost function combining two metrics: hops number and remaining energy making decide to select the fit route. Simulation results show that our protocol improved in terms of energy, network lifetime, and packet delivery ratio compared to the traditional protocols like DSR and AODV. In the future, we will focus on research and propose the security-aware routing protocols for the MANET-IoT networks.

References

1. C. K. Toh, A. Le and Y. Cho, "Load Balanced Routing Protocols for Ad Hoc Mobile Wireless Networks," IEEE Communications Magazine, vol. 47, no. 8, pp. 78–84, 2009.
2. Quy V. K., Nam V. H., Linh D. M., "A Survey of State-of-the-Art Energy Efficiency Routing Protocols for MANET," International Journal of Interactive Mobile Technologies, vol. 14, no. 9, pp. 215–226, 2020.
3. P. Yang, Y. Xiao, M. Xiao, S. Li., "6G Wireless Communications: Vision and Potential Techniques," IEEE Network, vol. 33, no. 4, pp. 70–75, 2019.

4. A. Ngoc Le and V. Minh Le, "Design a Simulation Model of Multi-radio Mobile Node in MANET," International Conference on Advanced Technologies for Communications (ATC), Nha Trang, Vietnam, pp. 237–240, 2020.
5. A. Dogra, R. K. Jha, S. Jain, "A Survey on beyond 5G Network with the Advent of 6G: Architecture and Emerging Technologies," IEEE Access (Early Access), 2020.
6. Hung L. N., Quy V. K, "A Review: Performance Improvement Routing Protocols for MANETs," Journal of Communications, vol. 15, no. 5, pp. 439–446, 2020.
7. S. Sen, J. Koo and S. Bagchi, "TRIFECTA: Security, Energy Efficiency, and Communication Capacity Comparison for Wireless IoT Devices," IEEE Internet Computing, vol. 22, no. 1, pp. 74–81, 2018.
8. Z. Niu, Q. Li, et al., "Identification of Critical Nodes for Enhanced Network Defense in MANET-IoT Networks," IEEE Access, vol. 8, pp. 183571–183582, 2020.
9. D. O. Akande et al., "A Network Lifetime Extension-Aware Cooperative MAC Protocol for MANETs With Optimized Power Control," IEEE Access, vol. 7, pp. 18546–18557, 2019.
10. Quy V. K., Ban N. T., Nam V. H., Tuan D. M., Han N. D., "Survey of Recent Routing Metrics and Protocols for Mobile Ad-Hoc Networks," Journal of Communications, vol. 14, no. 2, pp. 110–120, 2019.
11. RFC4728, https://www.ietf.org, accessed February 20, 2021.
12. RFC3561, https://www.ietf.org, accessed February 20, 2021.
13. K. K. Christopher, X. V. M. Arul and P. Karthikeyen, "Smart Toll Tax Automation and Monitoring System Using Android Application," 2019 IEEE International Conference on Intelligent Techniques in Control, Optimization and Signal Processing (INCOS), Tamilnadu, India, 2019, pp. 1–6, https://doi.org/10.1109/INCOS45849.2019.8951396.

Chapter 14
Enhanced Binary Krill Herd Algorithm for Effective Data Propagation in VANET

D. Saravanan, S. Janakiraman, Pon Harshavardhanan, S. Ananda Kumar, and D. Sathian

14.1 Introduction

Data propagation is a technique that guides the data transferring from the sender node to the receiver node using various techniques. It helps to deliver the data at the receiver end and helps end-to-end communication of sender and receiver. Here the data propagation technique is classified as Vehicle to Infrastructure/Infrastructure to Vehicle Propagation, Vehicle to Vehicle Propagation, Opportunistic Propagation, Geographical Propagation, Peer-to-Peer Propagation, Cluster-Based Propagation. VANET architecture can be broadly classified into three different categories or prototypes namely RVC, IVC, and HVC. In RVC, the vehicle has the tendency to transfer the data to the base station through roadside units. Since vehicles are fast-moving units, they cannot fit into a single topology for a standard period of time. This property leads to the handoff problem for the vehicles to receive any information either from RSU or from other vehicles [1, 2]. Another key factor to be considered is in case if the vehicle shares two different RSU ranges, there exists a problem to which RSU should receive the data. These problems can be addressed through RVC. In

D. Saravanan · P. Harshavardhanan (✉)
School of Computing Science and Engineering, VIT Bhopal University, Sehore, Madhya Pradesh, India
e-mail: pon.harshavardhanan@vitbhopal.ac.in

S. Janakiraman
Department of Banking Technology, Pondicherry University, Pondicherry, India

S. A. Kumar
School of Computer Science and Engineering, VIT University, Vellore, Tamil Nadu, India
e-mail: s.anandakumar@vit.ac.in

D. Sathian
School of Automation, Zhongkai University of Agriculture and Engineering, Guangzhou, China

S. Velliangiri et al. (eds.), *Secure Communication for 5G and IoT Networks*, EAI/Springer Innovations in Communication and Computing,
https://doi.org/10.1007/978-3-030-79766-9_14

IVC, the vehicles have the tendency to share the information without the presence of RSU via other vehicles. HVC is applicable for day-to-day life uses. In case a vehicle cannot access the RSU or any base stations, then it uses the other vehicles as routers and accesses the information [3, 4].

14.2 Literature Survey

Zhang, Xin, and Xiu Zhang [5] propose an effective approach for handling and providing an appropriate solution for spanning tree construction problems and this was achieved using a binary coded artificial bee colony (BABC) algorithm. The technique includes two important disparity techniques primarily calculated to uphold the consistency and with the maximum level of accuracy in binary coded solutions. The proposed algorithm is efficient to pact with a communication supported in roadside-to-vehicle communication. The success rate and average hitting time were well-thought-out as the two important aspects. Based on which it was discovered that MST was an influencing aspect which was also analyzed. It may be found that those BABC calculations might discover MST for 92% likelihood. However, it may be slower over Kruskal calculation as far as computational time, those BABC algorithms could accomplish a few suboptimal spanning trees clinched alongside a particular case run. This prescribes that those calculations might be helpful under the condition that three ways would oblige should be reconstructed habitually at the same time the system topology will be unaltered in a brief time [6–8].

Hanan, Abdul Hafidz Abdul, et al. [9] illustrate the theory behind the communication procedure carried in vehicular environments. It is purely depending on the two key factors such as real traffic data and investigates the advancement of location-centric dissemination of data. The current study is carried out with three key objectives (a) reveals proper data for carrying the analysis procedure with vibrant information about the traffic environment, in order to construct a proper data dissemination techniques; (b) the illustration of location-centric data dissemination in terms of functional and qualitative behavior of protocols, displayed with its pros and cons; and (c) to identify some upcoming research instructions in information dissemination and it also reveals the ideology behind the location. The traffic surroundings were categorized into three and it is based on the physical nature such as the speed of the vehicle, the distance between the inter-vehicular, the stability of the neighborhood, traffic density, and so on. The data traffic is measured for carrying the analyzed process on-road traffic environments based on the measurement of physical factors and weather circumstances [10, 11]. The major issues while constructing the network are identified by considering physical factors and weather circumstances placed for revealing the data dissemination. The characteristics of the functional and qualitative nature of location-centric techniques are discovered in view of urban and highway surroundings. IoT devices are entertained to establish the connection with the devices and propagate the data to the central unit. The relative analysis of the location-centric method is conceded for both urban and highway surroundings, and it is done independently depending on some exclusive and ordinary individuality of the surroundings [12, 13].

14.3 Problem Definition

The MST problem has been studied for decades; researchers still attempt to design an algorithm to tackle it in linear time. The problem is described as follows. Given an undirected graph $G = (N, E, W)$ where $W = \sum_{e \in E_T} w_e$ is weight function of defined on edge set E. A tree $T = (N, E_T, W_T)$ is called a spanning tree if it contains all nodes in graph G, where $W_T = \sum_{e \in E_T} w_e$ is the weight function of T. Among all spanning acyclic trees of G, the one having the least total weight is called minimum spanning tree of G (Fig. 14.1).

14.4 Enhanced Binary Krill Herd Algorithm for Effective Data Propagation

14.4.1 Angular Modulation Rule

Angle modulation (AM). This approach might have been utilized within the telecommunication industry for phase and frequency modulation of the signal. This system employs a trigonometric function that needs four parameters, and these parameters control the frequency and shift of the trigonometric function.

$g_i(x_j) = \text{sinsin}\,(2\pi\,(x_j - a_i)\,b_i \cos\,(2\pi(x_j - a_i)c_i)) + d_i$

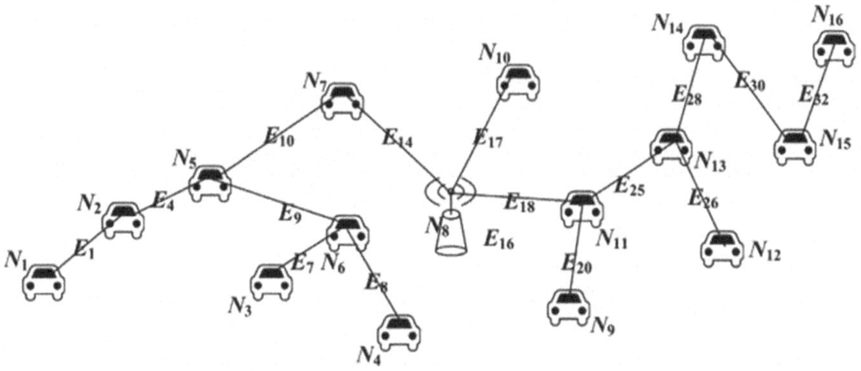

Fig. 14.1 Data ropagation through minimum spanning tree model

In binary heuristic optimization, this technique was initially applied in PSO using a set of benchmark functions. Consider an n-dimensional binary problem, and let $X = (x_1, x_2, \ldots, x_n)$ be a solution. We start with a four-dimensional search space, where each dimension represents a coefficient of (20). In the first step from four-dimensional space, we obtain a function in function space. Specifically, from every solution (a_i, b_i, c_i, and d_i) in this space, we obtained a g_i trigonometric function that lies in a function space.

Binarization. In the second step for each single element x_j, apply rule 24 and obtain an n-dimensional binary solution.

$b_{ij} = \{1,\ if\ g_i\ (x_j) \geq 0\ 0,$ otherwise

Algorithm 14.1: Binary Conversion
Input: Individual (x)

$\forall j \in n \mid j = \{1 \ldots n\}$

$g_i(x_j) = \text{sinsin}\ (2\pi\ (x_j - a_i)\ b_i \cos\ (2\pi(x_j - a_i)c_i)) + d_i$

$b_{ij} = \{1,\ if\ g_i\ (x_j) \geq 0\ 0,$ otherwise

End

Output: b

14.4.2 *Enhanced Binary Krill Herd Algorithm*

The proposed algorithm steps are clearly given in Algorithm 14.2. The proposed algorithm uses Binary conversion after the end of every iteration where the solution repairing process needs to be carried out. These solutions are first addressed with binary conversion and then the fitness of each solution will be identified [14, 15].

Variables Used

$d \leftarrow$ Problem Dimension

$\text{Max}_{\text{IT}} \leftarrow$ Total Number of Iterations

KHSize $\leftarrow$ Size of the Krill on each iteration

$\text{Pop}_{\text{KHSize} \times d} \leftarrow$ Possible solutions/Search space

$\text{KHIndiv}_{1 \times n} \leftarrow$ An individual Krill/Solution from the search space

Algorithm 14.2: Enhanced Binary Krill Herd for Data Dissemination (Pop, TC, G, n)

Step 1: Initialize Iter $\leftarrow 1$, $i \leftarrow 0$, $k \leftarrow 0$

Step 2: $\forall$KHIndiv $\in$ KHSize ***do*** *// Initialize the population using Chaotic Map*

$\text{Pop}_{\text{KHIndiv}} \leftarrow$ randi([0 1],) *// Generating Binary Values*

end for

Step 3: $\forall$*KHIndiv* $\in$ *KHSize* ***do***

$\text{Fit}(\text{Pop}_{\text{KHIndiv}}) \leftarrow f(\text{Pop}_{\text{KHIndiv}})$ *// Calling Objective Function*

end for

Step 4: Repeat through Step 9 **Until** $\text{Max}_{\text{IT}} \leq$ Iter, then go to Step 10

// Krill Herd Motion Process is done in three different forms

Step 5: Movement Induced by other Krills

Step 5.1: Repeat through Step 5.3 Until $i <$ KHSize | $i \in$ KHIndiv else goto Step 6

Step 5.2: $\alpha_i = \alpha_i^{\text{local}} + \alpha_i^{\text{target}}$ *// Position of each Krill can be induced by both local swarm density and global swarm density*

Step 5.3: $N_i^{\text{new}} = N^{\max}\ \alpha_i + \omega_n\ N_i^{\text{old}}$ *// Each individual in the Krill is induced by repulsive effect using* ω_n

Step 6: Foraging motion of individual Krills without Swarm colonial behavior

Step 6.1: Repeat through Step 5.3 Until $i <$ KHSize | $i \in$ KHIndiv else goto Step 7

Step 6.2: Powell Method ($\text{Pop}_{\text{KHIndiv}}$) *// Calling Powell Method*

Step 6.3: $\beta_i = \beta_i^{\text{food}} + \beta_i^{\text{best}}$ *// Its Previous best experience about the food location*

Step 6.4: $F_i = V_f \beta_i + \omega_f F_i^{\text{old}}$ *// The current update of Krill Foraging Motion with its foraging speed* V_f

Step 7: Individual movement of Krill Herd in a random manner

Step 7.1: Repeat through Step 7.2 Until $i <$ KHSize | $i \in$ KHIndiv else goto Step 8

Step 7.2: $D_i = D^{\max}\left(1 - \frac{I}{I_{\max}}\right)\delta$ *// The Actual Diffusion State of Each Krill*

// Krill Herd Motion Process on Position Update Level

Step 8: Repeat through Step 8.2 Until $i <$ KHSize | $i \in$ KHIndiv else goto Step 9

Step 8.1: $\frac{dX_i}{dt} = N_i + F_i + D_i$ *// Combine the information obtained through Krill herd Motion Process*

Step 8.2: $\Delta t = C_t \sum_{j=1}^{NV} \left(\text{UB}_j - \text{LB}_j\right)$ *// Compute the Limitation of each dimension*

Step 8.3: $X_i\left(t + \Delta t\right) = X_i\left(t\right) + \Delta t \frac{dX_i}{dt}$ *// Final Position Update of Each Krill*

Step 9: Applying Solution Repair and Binary Conversion Method

Step 7.1: Repeat through Step 9.2 Until $i <$ KHSize | $i \in$ KHIndiv else goto Step 10

Step 9.2: $\text{Pop}_{\text{KHIndiv}} \leftarrow$ BinConversion ($\text{Pop}_{\text{KHIndiv}}$) *// Calling Binary Conversion Function*

Step 9.3: Fit($\text{Pop}_{\text{KHIndiv}}$) $\leftarrow f$($\text{Pop}_{\text{KHIndiv}}$) *// Calling Objective Function*

Step 10: Return min(Fit($\text{Pop}_{\text{KHIndiv}}$))

14.5 Experimentation and Result Analysis

This section discusses the experimental setup and the experimental analysis carried out on the proposed algorithm.

Table 14.1 Experimental setup

Parameters	Values Small map	Medium map	Large map
Simulation time	3 min	3 min	3 min
Simulation area	1000 × 1000 m	1500 × 1500 m	2000 × 2000 m
Number of vehicles	10 Vehicles	20 Vehicles	30 Vehicles
Vehicles speed	0–50 km/h	0–50 km/h	0–50 km/h
Propagation model	Two ray ground	Two ray ground	Two ray ground
Radio frequency	2.47 GHz	2.47 GHz	2.47 GHz
Channel bandwidth	2 Mbps	2 Mbps	2 Mbps
Mac protocol	IEEE 802.11p	IEEE 802.11p	IEEE 802.11p
Transmission range	200 m	250 m	300 m
CBR data flow	5 Sessions	10 Sessions	15 Sessions

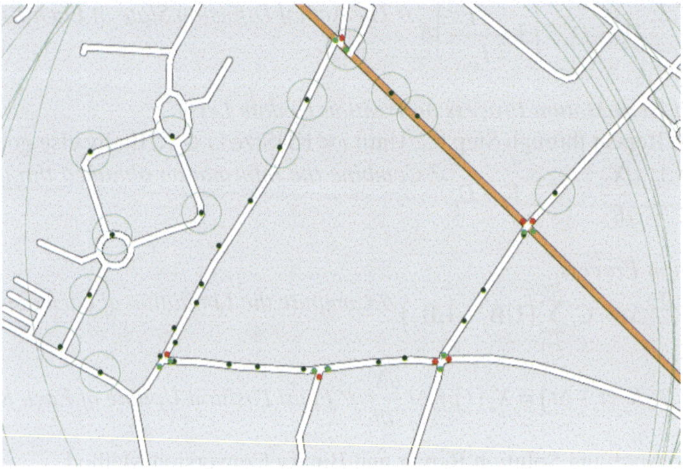

Fig. 14.2 Design I—small map (1000 × 1000 m)

14.5.1 Experimental Setup

For evaluation of the proposed algorithm, we have used the performance measures suchlike average delay and throughput. The results are tabulated in the respective tables, and the analysis of the results is given subsequently in graphical representation (Table 14.1, Figs. 14.2, 14.3, and 14.4).

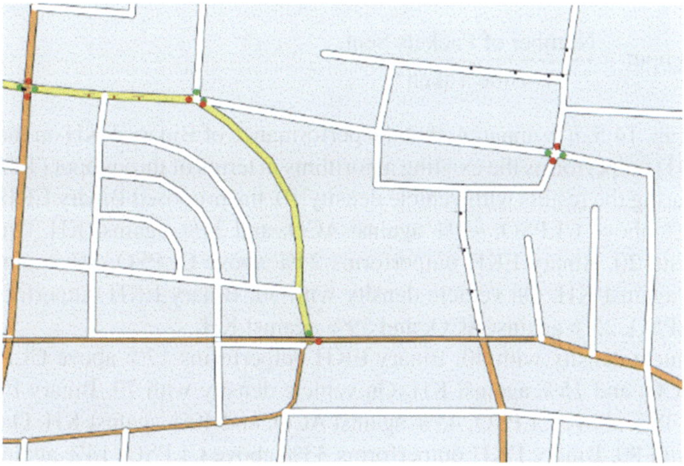

Fig. 14.3 Design II—medium map (1500 × 1500 m)

Fig. 14.4 Design III—large map (2000 × 2000 m)

14.5.2 Performance Metrics

Average End-to-End Delay

It is average time taken for a data packet to move from the source to the receivers

$$\text{Avg.EED} = \frac{\text{Total EED}}{\text{No.of Packets Sent}}$$

Throughput

Throughput refers to how much data can be transferred from the source to the receiver(s) in a given amount of time

$$\text{Throughput} = \frac{\text{Number of Packets Sent}}{\text{Time Taken}}$$

From Fig. 14.5, it is apparent that the performance of Binary-EKH on the Design Map-1 (D1) outperforms the existing algorithms in terms of throughput (Table 14.2). On comparing the results with vehicle density 10, the proposed Binary EKH outperforms 29% above CLPSO, −1% against ACO, and 37% against KH. On vehicle density with 20, Binary EKH outperforms 24% above CLPSO, 7% against ACO, and 19% against KH. On vehicle density with 30, Binary EKH outperforms 35% above CLPSO, 23% against ACO, and 39% against KH.

On vehicle density with 40, Binary EKH outperforms 17% above CLPSO, 6% against ACO, and 25% against KH. On vehicle density with 70, Binary EKH outperforms 48% above CLPSO, 45% against ACO, and 40% against KH. On vehicle density with 80, Binary EKH outperforms 33% above CLPSO, 14% against ACO, and 20% against KH. On vehicle density with 90, Binary EKH outperforms 50% above CLPSO, 31% against ACO, and 36% against KH. On vehicle density with 100, Binary EKH outperforms 41% above CLPSO, 41% against ACO, and 9% against KH.

From Fig. 14.6, it is apparent that the performance of Binary-EKH on the Design Map-2 (D2) outperforms the existing algorithms in terms of throughput. On comparing the results with vehicle density 10, the proposed Binary EKH outperforms 23% above CLPSO, 6% against ACO, and 9% against KH. On vehicle density with 20, Binary EKH outperforms 18% above CLPSO, 0% against ACO, and −3% against KH. On vehicle density with 30, Binary EKH outperforms 23% above CLPSO, 14% against ACO, and 12% against KH.

On vehicle density with 40, Binary EKH outperforms 21% above CLPSO, 21% against ACO, and 0% against KH. On vehicle density with 70, Binary EKH outperforms 21% above CLPSO, 9% against ACO, and 6% against KH. On vehicle density with 80, Binary EKH outperforms 31% above CLPSO, 16% against ACO, and 8% against KH. On vehicle density with 90, Binary EKH outperforms 20% above CLPSO, 22% against ACO, and 15% against KH. On vehicle density with 100,

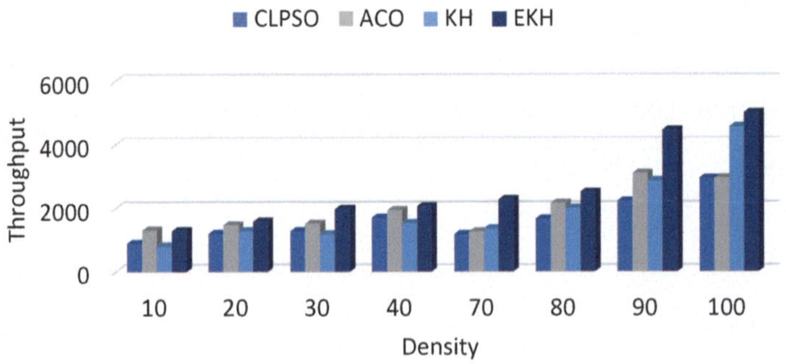

Fig. 14.5 Throughput rate of EKH on map D1 w.r.t. different number density

Table 14.2 Experimental results of EKH on different density of Vehicles w.r.t. Throughput Rate

	CLPSO [4]			ACO [5]			KH [6]			EKH		
Density	D1	D2	D3	D1	D2	D3	D1	D2	D3	D1	D2	D3
Low	906.36	386.55	386.55	1301.3	468.03	368.18	810.63	456.7	408.68	1284	499.84	409.01
	1207.3	514.56	514.56	1467.7	623.35	479.65	1282.4	644.09	529.45	1579.9	624.79	573.35
	1282.7	660.33	660.33	1510	736.21	597.77	1196.4	757.01	677.8	1971.7	855.75	695.4
	1711	811.77	811.77	1923.6	812.38	702.25	1541	1026.9	792.11	2050.4	1027	993.81
High	1180.7	1057.2	1057.2	1253.7	1216.8	1318.7	1361.3	1253.6	1223.4	2268.4	1337.8	1472.7
	1668.8	1502.6	1502.6	2146.3	1821	1878.2	2003.5	2003.7	1878.8	2502.4	2167.7	2177.4
	2227	2290.7	2290.7	3081.4	2225.3	2329.7	2861.7	2428.4	2634	4442.6	2860.8	2635.2
	2943.8	2566.2	2566.2	2943.4	2943.9	3053	4548.1	3033.6	3293.7	5004.7	3332.8	3910.5

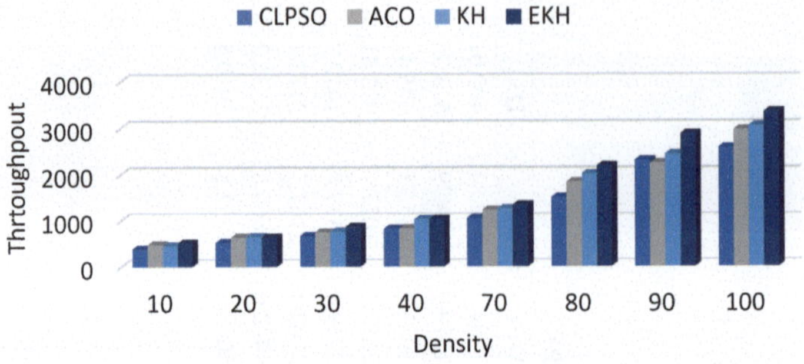

Fig. 14.6 Throughput rate of EKH on map D2 w.r.t. different number density

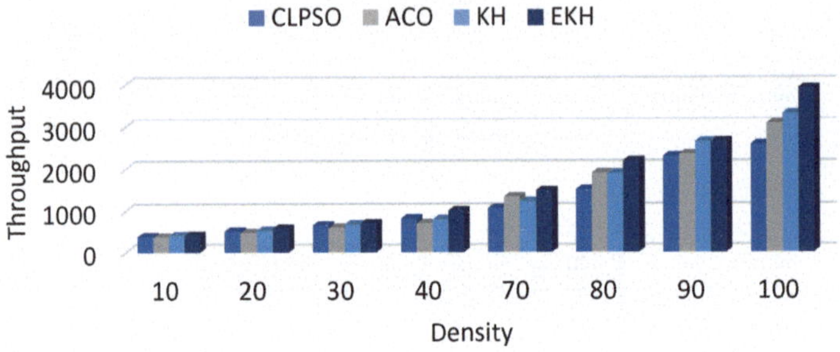

Fig. 14.7 Throughput rate of EKH on map D3 w.r.t. different number density

Binary EKH outperforms 23% above CLPSO, 12% against ACO, and 9% against KH.

From Fig. 14.7, it is apparent that the performance of Binary-EKH on the Design Map-3 (D3) outperforms the existing algorithms in terms of throughput. On comparing the results with vehicle density 10, the proposed Binary EKH outperforms 5% above CLPSO, 10% against ACO, and 0% against KH. On vehicle density with 20, Binary EKH outperforms 10% above CLPSO, 16% against ACO, and 8% against KH. On vehicle density with 30, Binary EKH outperforms 5% above CLPSO, 14% against ACO, and 3% against KH.

On vehicle density with 40, Binary EKH outperforms 18% above CLPSO, 29% against ACO, and 20% against KH. On vehicle density with 70, Binary EKH outperforms 28% above CLPSO, 10% against ACO, and 17% against KH. On vehicle density with 80, Binary EKH outperforms 31% above CLPSO, 14% against ACO, and 14% against KH. On vehicle density with 90, Binary EKH outperforms 13% above CLPSO, 12% against ACO, and 0% against KH. On vehicle density with 100, Binary EKH outperforms 34% above CLPSO, 22% against ACO, and 16% against KH (Table 14.3).

From Fig. 14.8, it is apparent that the performance of Binary-EKH on the Design Map-1 (D1) outperforms the existing algorithms in terms of average time delay. On comparing the results with vehicle density 10, the proposed Binary EKH

Table 14.3 Experimental results of EKH on different density of vehicles w.r.t. average delay

	CLPSO			ACO			KH			EKH		
Density	D1	D2	D3	D1	D2	D3	D1	D2	D3	D1	D2	D3
Low	425.54	2633	4041.8	416.16	2480.2	3652	527.01	2445.2	3461.1	341.84	2038.3	3228.6
	660.88	3275.7	4254.5	611.62	2757.2	4307	488.52	2824.8	3995.6	463.36	2344.8	3511.6
	1138.2	3394.4	4653.3	815.54	3224.5	4326.4	842.31	3389.1	4220.5	633.68	2753.8	3895.3
	1615	3637.8	4935.6	1387.8	3528.3	4823.8	1142.8	3270.8	4582.1	964.33	3010.4	4211.7
High	846.52	3876.3	5931.6	790.41	3628.4	5860.6	617.55	3412.8	5787.4	518.81	3119	5670.9
	960.56	4112.4	6584.4	1046.6	4192.6	6701.4	1015.3	3535	6188.1	734.28	3402.4	5856.5
	1374.8	4903.5	6863.9	1167.3	4359.1	6620.3	1042.5	4068.9	6800.9	952.17	3779.8	6127.5
	1544.1	5265.5	7337.5	2120	4330.8	7226.7	1705.9	4068.1	6929.3	1197.5	3935.6	6578.4

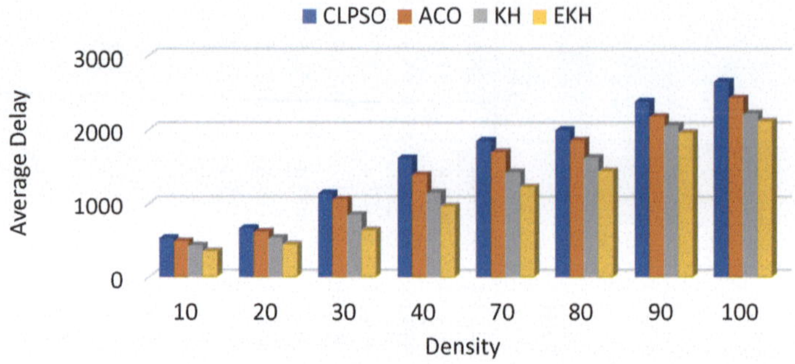

Fig. 14.8 Average delay of EKH on map D1 w.r.t. different number density

outperforms 35% above CLPSO, 29% against ACO, and 19% against KH. On vehicle density with 20, Binary EKH outperforms 34% above CLPSO, 29% against ACO, and 17% against KH. On vehicle density with 30, Binary EKH outperforms 44% above CLPSO, 40% against ACO, and 25% against KH.

On vehicle density with 40, Binary EKH outperforms 40% above CLPSO, 31% against ACO, and 16% against KH. On vehicle density with 70, Binary EKH outperforms 34% above CLPSO, 28% against ACO, and 14% against KH. On vehicle density with 80, Binary EKH outperforms 28% above CLPSO, 22% against ACO, and 11% against KH. On vehicle density with 90, Binary EKH outperforms 18% above CLPSO, 10% against ACO, and 4% against KH. On vehicle density with 100, Binary EKH outperforms 20% above CLPSO, 13% against ACO, and 5% against KH.

From Fig. 14.9, it is apparent that the performance of Binary-EKH on the Design Map-2 (D2) outperforms the existing algorithms in terms of average time delay. On comparing the results with vehicle density 10, the proposed Binary EKH outperforms 23% above CLPSO, 18% against ACO, and 17% against KH. On vehicle density with 20, Binary EKH outperforms 28% above CLPSO, 15% against ACO, and 17% against KH. On vehicle density with 30, Binary EKH outperforms 19% above CLPSO, 15% against ACO, and 19% against KH.

On vehicle density with 40, Binary EKH outperforms 17% above CLPSO, 15% against ACO, and 8% against KH. On vehicle density with 70, Binary EKH outperforms 20% above CLPSO, 14% against ACO, and 9% against KH. On vehicle density with 80, Binary EKH outperforms 17% above CLPSO, 19% against ACO, and 4% against KH. On vehicle density with 90, Binary EKH outperforms 23% above CLPSO, 13% against ACO, and 7% against KH. On vehicle density with 100, Binary EKH outperforms 25% above CLPSO, 9% against ACO, and 3% against KH.

From Fig. 14.10, it is apparent that the performance of Binary-EKH on the Design Map-3 (D3) outperforms the existing algorithms in terms of average time delay. On comparing the results with vehicle density 10, the proposed Binary EKH outperforms 20% above CLPSO, 12% against ACO, and 7% against KH. On vehicle density with 20, Binary EKH outperforms 17% above CLPSO, 18% against

ACO, and 12% against KH. On vehicle density with 30, Binary EKH outperforms 16% above CLPSO, 10% against ACO, and 8% against KH.

On vehicle density with 40, Binary EKH outperforms 15% above CLPSO, 13% against ACO, and 8% against KH. On vehicle density with 70, Binary EKH outperforms 4% above CLPSO, 3% against ACO, and 2% against KH. On vehicle density with 80, Binary EKH outperforms 11% above CLPSO, 13% against ACO, and 5% against KH. On vehicle density with 90, Binary EKH outperforms 11% above CLPSO, 7% against ACO, and 10% against KH. On vehicle density with 100, Binary EKH outperforms 10% above CLPSO, 9% against ACO, and 5% against KH.

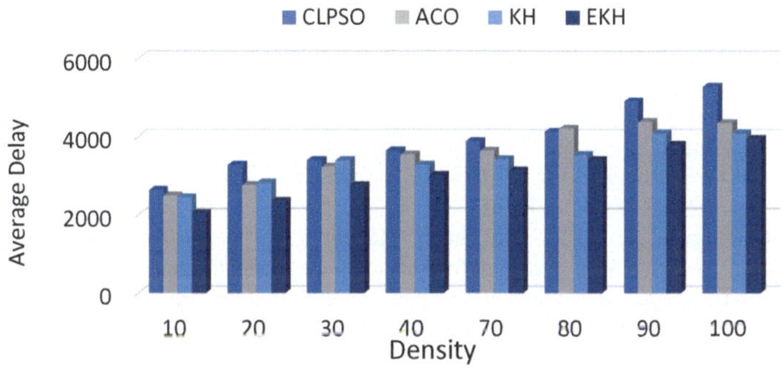

Fig. 14.9 Average delay of EKH on map D2 w.r.t. different number density

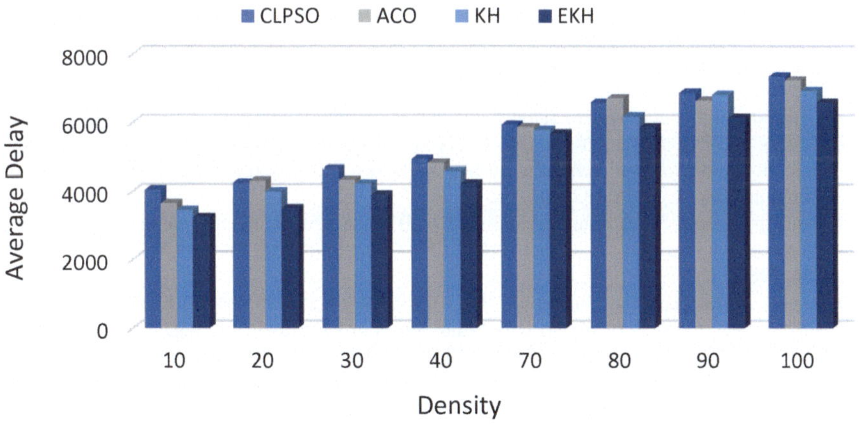

Fig. 14.10 Average delay of EKH on map D3 w.r.t. different number density

14.6 Conclusion

In this paper, EKH algorithm is used to solve data propagation in VANET by structuring an effective minimum spanning tree model. For representing the solutions so as to solve data propagation through a minimum spanning tree, the solutions are given in terms of binary using angular modulation rule which is a measurable mathematical model. On every iteration, the angular modulation rule takes care of the continuous variables and converts it into binary form. The results of the proposed algorithm are evaluated using standard performance measures and compared with the recent algorithms to show the significance of the proposed mechanism.

References

1. Aslam, B., Amjad, F., & Zou, C. C. (2012, July). Optimal roadside unit's placement in urban areas for vehicular networks. In 2012 IEEE Symposium on Computers and Communications (ISCC) (pp. 000423–000429). IEEE.
2. K. K. Christopher, X. V. M. Arul and P. Karthikeyen, "Smart Toll Tax Automation and Monitoring System Using Android Application," 2019 IEEE International Conference on Intelligent Techniques in Control, Optimization and Signal Processing (INCOS), Tamilnadu, India, 2019, pp. 1–6, https://doi.org/10.1109/INCOS45849.2019.8951396.
3. Liang, J. J., Qin, A. K., Suganthan, P. N., & Baskar, S. (2006). Comprehensive learning particle swarm optimizer for global optimization of multimodal functions. *IEEE transactions on evolutionary computation*, *10*(3), 281–295.
4. Zhang, Xin, and Xiu Zhang. "A binary artificial bee colony algorithm for constructing spanning trees in vehicular ad hoc networks." Ad Hoc Networks 58 (2017): 198–204.
5. Dorigo, M., & Di Caro, G. (1999, July). Ant colony optimization: a new meta-heuristic. In *Proceedings of the 1999 congress on evolutionary computation-CEC99 (Cat. No. 99TH8406)* (Vol. 2, pp. 1470–1477). IEEE.
6. Gandomi, A. H., & Alavi, A. H. (2012). Krill herd: a new bio-inspired optimization algorithm. *Communications in nonlinear science and numerical simulation*, *17*(12), 4831–4845.
7. Zhuang, Y., Pan, J., Luo, Y., & Cai, L. (2011). Time and location-critical emergency message dissemination for vehicular ad-hoc networks. IEEE Journal on Selected Areas in Communications, 29(1), 187–196.
8. Panichpapiboon, S., & Pattara-Atikom, W. (2012). A review of information dissemination protocols for vehicular ad hoc networks. IEEE Communications Surveys & Tutorials, 14(3), 784–798.
9. Hanan, Abdul Hafidz Abdul, et al. "Real traffic-data based evaluation of vehicular traffic environment and state-of-the-art with future issues in location-centric data dissemination for VANETs." Digital Communications and Networks 3.3 (2017): 195–210.
10. Yan, T., Zhang, W., Wang, G., & Zhang, Y. (2014). Access points planning in urban area for data dissemination to drivers. IEEE Transactions on Vehicular Technology, 63(1), 390–402.
11. Chaqfeh, Moumena, Abderrahmane Lakas, and Imad Jawhar. "A survey on data dissemination in vehicular ad hoc networks." *Vehicular Communications* 1.4 (2014): 214–225.
12. Liu, Xiaoyun, and Gongjun Yan. "Analytically modeling data dissemination in vehicular ad hoc networks." *Ad Hoc Networks* 52 (2016): 17–27.
13. Zhu, Junyu, et al. "EDDA: An efficient distributed data replication algorithm in VANETs." *Sensors* 18.2 (2018): 547.

14. Iza-Paredes, Cristhian, et al. "Game-theoretical design of an adaptive distributed dissemination protocol for VANETs." *Sensors* 18.1 (2018): 294.
15. V. Anbarasu, P. Karthikeyan and S. P. Anandaraj, "Turning Human and Food Waste into Reusable Energy in a Multilevel Apartment Using IoT," 2020 6th International Conference on Advanced Computing and Communication Systems (ICACCS), Coimbatore, India, 2020, pp. 440–444, doi: https://doi.org/10.1109/ICACCS48705.2020.9074170.

Index

S. Velliangiri et al. (eds.), *Secure Communication for 5G and IoT Networks*, EAI/Springer Innovations in Communication and Computing,
https://doi.org/10.1007/978-3-030-79766-9

K

The manufacturer's authorised representative in the EU is Springer Nature Customer Service Centre GmbH, Europaplatz 3, 69115 Heidelberg, Germany. If you have any concerns regarding our products, please contact ProductSafety@springernature.com

Printed and bound by CPI Group (UK) Ltd, Croydon, CR0 4YY
30/09/2026
02225694-0006